THE SUNDAY TELEGRAPH COOKERY BOOK

BY JEAN ROBERTSON
With the chapter on wine by ANDREW ROBERTSON

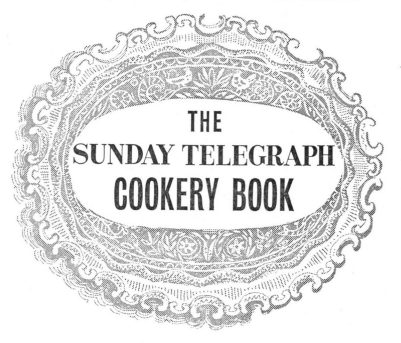

THE
SUNDAY TELEGRAPH
COOKERY BOOK

Illustrated by PAMELA GUILLE
Colour photographs by BRIAN PRICE THOMAS
also PENNY TWEEDIE

COLLINS
for
THE COOKERY BOOK CLUB

© 1965

A Sunday Telegraph Publication

This edition published 1967 by
THE COOKERY BOOK CLUB
9 Grape Street
London, W.C.2
for sale only to its members

Printed in Great Britain by Collins Clear-Type Press

*Colour plates printed in the Netherlands
by Jon Enchedé en Zonen, Haarlem*

I WOULD LIKE TO THANK...

MR. BOWEN, *the buyer in Harrods Provision Department, for giving up his time to teach me something of the subtle British craft of ham curing, and for providing the hams for the photograph facing page 161.*

HABITAT, *and* MISS SASSON *of the basement in particular, who lent me almost all the kitchenware pictured in the kitchenscape facing page 64*

SENHOR FERNANDO GUEDES *of Oporto who sent me the remarkable slatted dish in the foreground of the kitchenware picture. It is used in Portugal to scent meat and* chourica *sausages in particular in the fumes of flaming brandy, it is also ideal for perfuming fish with the scent released by the burning stalks of dried herbs. English potters, please copy.*

FORTNUM AND MASON, *for kindly lending us some of the jars of preserves for the photograph facing page 240.*

MR. HUGMAN, *Managing Director of the Whitstable Oyster Fishery. The month we needed oysters in order to take the picture had no R in it and every oyster in the British Isles was sleeping peacefully in its sea-bed. Nothing daunted, Mr. Hugman donned his diving gear and went down to the beds and fetched up a dozen fine fat "natives". The photographic session was followed by an oyster feast, so Mr. Hugman earned our gratitude twice over.*

MR. DAVID PEPPERCORN, *Master of Wine and joint Managing Director of Fosters Wine Merchants, for reading my husband's chapter on wine.*

7

8 *I Would Like to Thank* . . .

MRS. ELIZABETH DAVID *for practical help in the form of the tripière in the tripe picture facing page 144. Also, and far more important, for her inspiration. It was Elizabeth David with her scholarly integrity and magnetic style who first made me want to cook. Her influence will be manifest to anyone who uses this book.*

MISS EILEEN WYNNE *of the* Sunday Telegraph *who somehow found the time, amid a myriad other tasks, to compile the files on which this book is based.*

MISS IRENA BESASPARIS, *whose enthusiasm, wit and sheer dogged hard work helped me to turn that pile of files into this book.*

MISS WINEFRIDE JACKSON, *Woman's Editor of the* Sunday Telegraph, *for giving me the opportunity, the incentive and the time to write this book, and for her constant encouragement.*

I am also indebted to the TIMES PUBLISHING COMPANY *for permission to reprint four of my recipes which first appeared in* The Times, MACDONALD AND COMPANY LIMITED *for permission to use material from Dorothy Hartley's* Food in England, *and from* Italian Food *by Elizabeth David,* FABER AND FABER LIMITED *for permission to reprint the recipe from* Greek Cooking *by Joyce Stubbs,* JONATHAN CAPE LIMITED *for permission to reprint the recipe from* Good Things in England *by Florence White,* BRUNO CASSIRER LIMITED *for permission to reprint the recipe from* Cooking with Pomiane *by Edouard de Pomiane,* VIKING PRESS LIMITED *for permission to reprint the recipe from* Casserole Cooking *by Marie Tracy.*

The photographs of Pies *and* Pigeons *are reproduced by courtesy of* THE DAILY TELEGRAPH WEEKEND MAGAZINE.

–1–
TIME
AND THE COOK

This book is a Kitchen Companion for enthusiasts who enjoy cooking
as much as eating.

Eating is one of life's daily pleasures — but only one of them;
and cooking, vital as it is to the contentment and well being of the
people she cares for is only one of the preoccupations of the complete
woman. Her delight in it is likely to depend as much on her ability
to control and vary the task of producing daily meals as it does on
her skill.

One of the keys to limiting the potentially limitless task of feeding
the family (and friends) is concentration — not of the mind, but of
the cooking and its inexorable by-product, washing up.

Concentrate the cooking into a couple of serious sessions a week,
putting together generous casserole dishes (*see Casserole chapter,
pp. 89–115*) which can appear a second time without any extra trouble,
roasting meat and poultry in ways that make it as good to eat cold
as hot (*see Cold Food chapter, pp. 133–41*) — interspersed, of course,
with meals fresh from the oven or the grill. Make dishes capable of
playing several parts, and basic preparations, such as *roux* or syrup,
in double quantities — half for today and half for storing away in
the fridge for the day after the day after tomorrow.

So many foods and dishes lend themselves to instalment cooking of this sort and it rarely takes more than time and a quarter, and often no more time at all, to produce double quantities.

For example :

- with *mayonnaise*, it is getting the emulsion going that takes the time, so why not make half a pint while you are at it

- the only work attached to boiling rice is washing up the pan afterwards, but two cups won't make that pan any stickier than one, and rice keeps well for two or three days in a fridge box, and heats up easily

- if you have to chop an onion, chop 2, and store the spare one under oil in the fridge — it won't take any longer to get the smell off your hands and the chopping board.

It seems almost profligate not to make *ratatouille* by the potful; it has so many roles and keeps so well (being stewed in oil).   It can :

- appear cold, sprinkled with herbs, as a starter

- replace gravy and any other sauce with lamb

- serve as a bed for *oeufs sur le plat* or as a filling for omelettes

- accompany coarse fish

- make a *moussaka* — with lamb cooked or raw.   (This is the nicest way I know of finishing up the remains of lamb or mutton.)

Pâté stays good for weeks in the fridge if it is properly sealed with pure melted lard, so why not make two or even three when the *pâté*-making mood comes upon you?

With ruses such as these (and there are many, many more) and just a little programming, I reckon that a well-organised woman should be able to produce at least two of the week's meals in 15 minutes flat.

This means, of course, scrapping the old strait-jacket of meat and 2 veg. followed by pudding, and concentrating on 1 *plat du jour*, accompanied by rice or egg noodles (or some good bread and butter), a salad, fruit and cheese.   Puddings are part of our way of eating in this country (they've got the second largest chapter in this book),but this does not mean that they have to be a fixture on every menu and cannot play box and cox with fruit and cheese.

It also means making intelligent use of prepared foods, buying those which the professionals do well, such as smoked meat and fish, some *pâtés*, frozen spinach and puff pastry, in order to have the time to make those which professionals do badly (ice cream and marmalade, for instance). A good working rule for a busy woman with guests on her hands is to buy the first course, casserole the second and make the third — or make do without one.

Sailing an honest course between the lure of convenience eating and the healthy appeal of fresh, unsophisticated food is one of the dilemmas of modern housekeeping. It isn't difficult to convince oneself that anything cooked at home is better than anything that ever came out of a factory. It invariably is. But we can't all afford the luxury of having the best all the time, and the choice is a matter of priorities — which is more important, a bedtime story or freshly podded peas for dinner, that a woman should have the satisfaction of an interest outside the family, or all the baking should be done at home ? Guilt plays some part in the choice, so does vanity : guilt engendered by the feeling that anything too easy must somehow be sinful; vanity because some of us hate to have to admit to our guests that anything on the table is not a product of our own skill. Thus conscience can make galley slaves of us all. It's a hard course to chart, and one of the aims of this book is to put out a few markers.

-2-
THE KITCHEN

• THE SOURCE OF HEAT •

Personal preference is the guiding factor in making the choice, which is, in effect, between gas and electricity. Unfortunately, the ample comfort and perpetual heat of solid fuel cookers is hardly practical in the clinical cubicles that architects call kitchens these days.

Both gas and electricity have their champions and both now have most of the virtues which were once the monopoly of the other. Electric hobs have become quicker, more sensitive and easier to clean than seemed possible 10 years ago; while gas is cleaner and safer, and gas cookers now come equipped with many of the features which were once exclusive to electricity — such as thermostatically controlled hobs and time controlled oven switches, though few of them have the elegance of the electric cookers.

My own preference is for gas, which I find still has the edge over electricity as a cheap and obedient kitchen servant. Electricity cannot emulate gas in responding instantly to the touch of a tap. However, that is a personal view.

What matters more is that householders should not be deprived of the right to make their personal choice in this matter by builders, property developers or local authorities who decide for reasons of economy (theirs, not the householder's) not to run gas pipes to new housing estates and flat blocks.

Points to bear in mind when you go shopping for a cooker

• Try, if possible, to anticipate your requirements five or even ten years from now when both your family and your ambitions as a cook may have outgrown the cooker that suited you in your first year of marriage. Nothing makes more of a chore out of cooking than a cramped little stove with one burner too few and a titchy little oven that can only take one dish at a time (small ovens, incidentally, get dirty quicker than big ones). But as there is no effective market for secondhand cookers, a new one is an unlikely luxury for most families.

• Remember that you pay for every refinement on a cooker and it is a waste of money to pay for those you are never likely to use. A time switch in the oven is a boon to the wife out at work, but to the woman at home all day just an unemployed extra.

• If you own your own home, or have a long lease, consider very seriously split-level cooking before opting for a conventional free-standing stove. Split level costs more, and it is more trouble to install, but it adds immeasurably to the comfort, convenience and safety of the kitchen. The oven is out of the reach of toddlers but in comfortable reach of the cook. With spinal disorders becoming one of our contemporary scourges (and greatly aggravated by constant stooping) this is a factor which cannot be lightly dismissed.

• PANS AND POTS •

PANS

Saucepans come into the shrinking category of goods where fashion does not count but quality does. It still pays to save up for the very best pans you can possibly afford. There is no such thing as a good cheap saucepan, though there are some pretty poor expensive pans to be found in the shops.

The two most satisfactory materials for saucepans (excluding copper, a solid luxury few can afford) are heavy gauge aluminium and cast-iron. Both metals are good conductors of heat (which means they spread it quickly and evenly), both are easily cared for and strong, and both have had their one major defect eliminated by the technicians.

The slight reaction between an aluminium pan and its contents worried some people. Crown Merton have recently disposed of this problem by putting a lining of stainless steel, which is totally inert, into one of their aluminium ranges (which they call "Stainless Steel Plus"). This being a highly competitive world, their sensible lead is likely to be followed shortly by other aluminium pan manufacturers.

Cast-iron in its untreated state has a more serious shortcoming as a metal for cooking in — it reacts strongly to the acid in food, and it rusts. The solution, not a new one, is a coating of vitreous enamel. But it is only relatively recently that *almost* chip-proof vitrified cast-iron pans have been widely distributed in Britain. One of the most reliable makes is Le Creuset which are imported from France by Clarbat Ltd. (302 Barrington Road, London, S.W.9).

Points to consider when buying saucepans

• You need one really large pan (at least 5 quarts) for *pot au feu* and for boiling chickens, ham, rice and *pasta*, but this is the one pan that does not have to be expensively solid.

• You need one small heavy pan with an insulated handle for sauces.

• With your large and medium sized pans it can be an advantage to have a pair of small ear-like metal handles instead of one long one. They are safer when there are small children around, also they enable the pan to go from the hob to the oven. On the other hand, eared pans need two hands to lift them. It's a moot point, but worth considering.

• A *sauté* pan, a shallow straight-sided saucepan with a lid, is one of the most useful vessels in the kitchen. It should be your first luxury after acquiring a basic set (3, 5 and 7 pint) of conventional saucepans.

• If you are going to treat yourself to something in copper, make it a frying pan and make sure it is a heavy one. Heaviness in a frying pan is even more important than in a saucepan. Lightweight copper is suitable only for chafing dish cookery, or hanging on the kitchen wall. Copper cooking utensils are widely sold these days, but if you want advice on the best pan for your particular purpose go to a professional kitchen shop such as Jaeggi (232 Tottenham Court Road) or a store with a serious kitchen department such as Heals (also in Tottenham Court Road) or The General Trading Company (144 Sloane Street, S.W.1).

• No domestic grill that I have seen gets hot enough to grill steak properly, which makes a top-of-the-stove grill, a kind of ridged frying pan made of cast-iron, a worthwhile indulgence for steak fanciers.

• Until you have money to spare, a bowl resting on the rim of a saucepan will serve as a double saucepan (or porringer).

• Make sure lids fit snugly, and that rivets holding the handles in place lie flush to the inside of the pan. If they don't, they make tiresome food traps.

POTS

The cooks of the 60's have rediscovered the gentle dependability of earthenware, which is as old as the art of cooking. It is both cheap and dignified, it is slow to absorb heat and equally slow to lose it again, and it is surprisingly robust. Sitting on a wire or asbestos mat it can withstand the direct heat of gas (but not electricity), and it goes unselfconsciously to the dining room table where it will keep its contents hot longer than any other material. A big lidded casserole, a tall stew pot of 1 gallon capacity, a *terrine* for *pâtés*, a round "saucepan" with one handle and a couple of oval *gratin* dishes — any or all of these in earthenware would be a blessing in anyone's kitchen.

Pearson of Chesterfield make a remarkable range of serviceable earthenware pots of all sizes and shapes which is so cheap that some retailers will not give it shop room. However, all the more enlightened kitchen shops stock Pearson's admirable wares, and a number (such as Woollands and John Lewis) also have the handsome French earthenware.

• CUTTING AND CHOPPING •

KNIVES

The knife is the most important tool in the kitchen, and a little battery of carbon steel knives, all with edges as sharp as a jester's wits, are worth a drawerful of ingenious gadgets.

Stainless steel is no good for most kitchen jobs, though one wavy-edged stainless steel knife is needed for cutting bread, fruit and acid vegetables. Carbon steel does stain, of course, but the tiny task of cleaning the knives is a small price to pay for the convenience of having blades that can be made almost lethally sharp with a few strokes of a carborundum stone.

Sound, well-balanced French cook's knives (the best buy for serious cooks) have wide, strong shoulders and blades that taper upwards towards the point so that the knife rocks as it cuts. The best knives also have riveted handles and bolstered blades (that is a blade which runs right through the width and breadth of the handle).

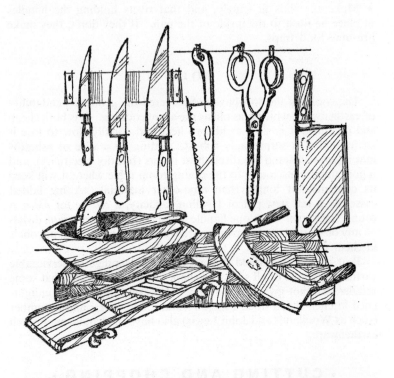

If there are any cook's knives finer than the black handled Sabatier I have yet to discover them. However, Sheffield is at last beginning to take the domestic cook seriously, and Joseph Rodgers' workmanlike knives, which are cheaper than Sabatier, can be found in good quality kitchen shops.

Three French cook's knives, large (9-inch), medium (6-inch) and tiny (3-inch) is a good basic armoury.

If you do invest in serious kitchen knives, spare them the rough and tumble of life in the kitchen drawer. Buy a magnetic knife rack and screw it to the wall. It will hold all your knives in the isolation they need, but always ready to hand.

OTHER USEFUL BLADES

• The fastest and most efficient chopper I know of is a double-handled knife with a half-moon shaped blade called, appropriately enough, a *mezzaluna* and costing 16s. or 17s. It works with a see-saw motion and rapidly reduces herbs, vegetables, cooked meat and nuts to a fine hash. It does most of the work of a mincer and it is so much easier to wash up.

• The *Mandoline* (or Universal slicer as it is called in England) is an adjustable wooden frame equipped with a trio of blades. It makes short work of all slicing and "matchstick" jobs; for instance, a whole cucumber can be reduced to transparent slices in 2 minutes flat. Price — between 20s. and 25s.

• A kitchen saw (Tala make an efficient one for about 5s.) is invaluable for tackling intractable joints and severing bones.

• A small cleaver — the sharp edge for big bones, the flat blade for flattening and tenderising.

• A swivel-bladed potato peeler (stainless steel is best) which will take the zest off an orange, the rind off a cheese, the skin off a pear, as well as the peel off a potato.

• A pair of kitchen scissors with one serrated blade.

• KITCHEN TOOLS •

Once you have acquired the bare bones of a saucepan wardrobe and one or two sharp knives, it is remarkable how much you can do without. Undoubtedly the right pot contributes to the authenticity of a dish and simplifies the work of the cook. But it is not indispensable. Even a *soufflé* can be made in a round earthenware casserole in an emergency. Never be put off a recipe because you haven't got the utensil or gadget it mentions. *In this book, anyway, the equipment mentioned in recipes is a counsel of perfection‘ not a prerequisite.*

On the other hand there are a few devices without which no cook can work efficiently, by which I mean simple tasks take twice as long as they need. Most of these cost less than £1 and some less than 5s.

Indispensable to any well run kitchen

A ball of thin string
A yard of muslin (for making spice bags or holding anything, such as pips or bones which have to be removed from the finished dish)
Wooden spoons
A glass funnel (plastic ones are almost impossible to keep clean)
Wire or asbestos mats
A straining spoon
A large ladle
A conical sieve in stainless steel for straining sauces or gravies
Two conventional sieves (one at least 7 inches wide, and one made of nylon for the sieving and straining of soup)
A large wooden "mushroom" for pressing food through the sieve
A flour dredger
A palette knife
A flexible rubber scraper for getting the last scrap out of bowls and saucepans, and cleaning up the plates after dinner
A hooped wire whisk for whipping egg whites and blending mixtures that would be all the better for a breath of air
Metal tongs for turning food in a frying pan or under the grill, or for lifting large objects such as fruit jars or artichokes out of boiling water
Two narrow paint brushes (one for painting meat and the other fish with oil before frying, grilling or roasting)
A pepper grinder
A "Mouli" cheese mill (which doubles as a nut grinder)
A "Mouli" vegetable mill which sieves as it strains anything sievable
A heavy metal baking sheet (this usually comes with the cooker)
A couple of flan rings
A collapsible wire basket for lettuce drying, deep frying, insulating, draining
A pie dish
A set of measuring spoons and a British Standard measuring cup (*see* *p. 42*)
Plenty of paper — aluminium foil for cooking, wrapping, covering, etc.
　　　　　　　　 — absorbent for absorbing excess fat and water from food, as well as a million mopping-up jobs

And, finally, the most expensive essential, a large, thick, seamless chopping board.

Invaluable but not completely indispensable

A pestle and mortar (don't buy one that is too small)

A 1½-pint *soufflé* dish

A larding needle

An *hachinette*, a wooden chopping bowl with a fitting half-moon knife for herbs

A slab of marble for pastry

Egg dishes — 4 or 6 individual *ramekins* or *cocottes*

        — 4-6 flat-eared *oeufs sur le plat* dishes

An earthenware steamer (*cerafeu*) for vegetables.

• THE BLENDER •

There is room for more than one opinion on the usefulness of electric machines in the kitchen but not, I think, as far as the electric blender or liquidiser is concerned. Nearly every day I silently thank the genius who invented it. It takes up little room, it is easy to take apart, clean and reassemble, and its uses are legion. Apart from such obvious tasks as making a *purée* out of vegetables and fruit for soups and fools —

- it turns grains of every sort, such as coffee, lentils and chick peas, also nuts and sugar lumps, into a powder
- it takes the lumps out of sauces
- makes short work of stuffings
- turns bread, fresh or stale, into breadcrumbs (the only way of making breadcrumbs that is not tedious)
- emulsifies cooked left-overs such as vegetables, meat, fish for quick soups and spreads
- froths children's drinks — oh, those lovely milk shakes !
- chops ice
- makes whole orangeade and lemonade. Pare off the zest, remove pith, cut the fruit and put it with zest into the beaker with a little water and sugar (1 orange, 1½ gills water, 4 tsp sugar). For a smooth milky drink add ½ gill creamy milk.

- makes a quick *pâté* out of salmon, smoked or fresh. Put cream into the beaker, add chopped salmon and seasoning and switch on

- makes a cheese spread out of hard cheese, cream and seasoning.

Important : When emulsifying raw vegetables or hard fruit, there must always be liquid in the beaker. For a quick soup put some water or stock in the beaker and the chopped vegetables of your choice.

To clean : Half-fill with water and switch on.

• REFRIGERATION •

I reckon a refrigerator to be the fourth most important piece of equipment in the home (the first three places being filled by bed, cooker and "usual offices"). It is a year-round necessity which serves the housewife 24 hours a day — a vital factor in food hygiene and the starting point of rational housekeeping.

In its cold, silent way, a refrigerator cuts down time spent shopping and cooking, and money spent on wasted food. It enables the cook, among other things, to :

- buy large joints of meat rather than chops or cutlets
- buy at the cheapest time of the week (which often, contrary to general belief, is Saturday)
- make casserole dishes large enough to feed the family at least twice, and prepare or cook many things in double quantities, half for immediate use and half for the future
- buy large packs of frozen food or big tins which are better value than small ones.

Points to consider when choosing a refrigerator

• If you are not planning on investing in a deep-freeze cabinet, allow two cubic feet of storage space in the refrigerator for each member of the family, including possible new members. Fridges have a better secondhand value than cookers, but it still pays to buy one that will answer your needs for at least ten years.

• The star markings which appear on many refrigerators are not quality marks in the *Michelin* sense. They are indicators of the temperature of the freezing compartment and the length of time frozen food can safely be stored.

One star ✱ indicates a maximum temperature in the freezing compartment of 21 deg. F. and safe storage for 1 week.

Two stars ✱✱ indicates a maximum temperature in the freezing compartment of 10 deg. F. and safe storage for 1 month.

Three stars ✱✱✱ indicates a maximum temperature in the freezing compartment of 0 deg. F. and safe storage for 3 months.

• Unless you are planning on a frozen food cabinet, a large locker in the ordinary refrigerator is most desirable in this deep-frozen age.

• Not all defrosting devices that call themselves "automatic" or "push button" are as effortless to operate as the makers would have us believe. Completely automatic defrosting is still a rare attachment on British refrigerators.

• Distinguish between the *gross* cubic capacity of the cabinet and its *net* cubic capacity — a distinction that is not always made clear in the sales literature.

• Other points to look for :
Adjustable feet for uneven floors
Wheels so that you can clean behind the fridge
Removable egg rack: eggs should not be kept in the fridge if you have anywhere else to store them, so a fixed egg rack can be a waste of expensive space
Door racks: invaluable for all bits and pieces that tend to get lost in the body of the cabinet
Adjustable shelves

• Most refrigerators are lined with plastic these days. This has made them cheaper, lighter and less bulky, but it does make them more tenacious of persistent smells. There are still one or two machines made with a lining of vitreous enamel. It is arguable whether the advantages of plastic outweigh this one drawback.

• PLANNING THE KITCHEN FOR COMFORT •

The layout of the kitchen is something over which few of us have complete control. In modern flats and houses we have to live, and like it, in kitchens dreamed up by the builder or architect.

However, this need not preclude a few minor installations which can make all the difference between pleasure and frustration in the kitchen.

For instance, a couple of slatted shelves over the sink for saucepans can save any amount of rummaging in cupboards. It also prevents the saucepans from getting musty.

If your spices and herbs hang on a rack on the wall you are more likely to make use of them.

A long chest of storage drawers just one drawer deep hanging unobtrusively under a wall unit will hold all your regular groceries at the ready.

A peg board with hooks is useful for holding all manner of hangable gadgets and small tools at fingertip reach.

A magnetic rack for knives (already mentioned in the *Cutting and Chopping* section).

A waist level electric point for blender, coffee grinder, etc., saves having to lengthen the flex of everything you buy (some manufacturers, especially foreign ones, are so mean with flex.) It also saves constant bending.

A rack near the stove or sink, or both, for dispensing absorbent kitchen paper.

If you do have any say in the basic arrangement of your kitchen, bear in mind that :

• two sinks are more than twice as useful as one
• the stainlessness of stainless steel sinks and draining boards is a relative term
• every inch of width in a store cupboard is worth 4 of depth, deep cupboards in a kitchen are usually untidy and always an irritation.

-3-
THE CHARACTER
OF THE COOKING

· WINE IN THE KITCHEN ·

A kitchen without wine is like an orchestra without woodwind —
it lacks a dimension. Wine brings subtlety, fragrance and economy
to cooking; subtlety because in a properly prepared dish its absence
is more noticeable than its presence; fragrance because it is the
grapey residue which informs the dish, not the alcohol; and economy
because, judiciously used, wine tenderises and enriches the inferior
cuts and enables fine dishes to be produced from cheap meat.

Wine has many jobs in the kitchen :

• As a marinade (*see p. 39*). Wine used for marinading meat
can often be strained into the casserole in which the meat will be
cooked, or added at the last moment to the gravy.

• To enrich sauces and stews.

• To replace artificial thickenings and flavourings in gravy. The
simplest and most delicious gravy for roast or grilled meat is
made by adding a little wine, the colour varying with the colour of
the meat, to the juices in the drip pan. If there is a surplus of fat,

pour it off before adding the wine. Let the amalgam of wine and meat juices bubble fiercely. Add a few tablespoons of boiling water or good stock, bubble again and serve.)

• To enliven a soup. Add 1 tblsp of sherry to a pot of soup for four by way of seasoning at the last moment before serving.

• To ennoble a fine fruit, such as peach — bathe it in a little sweet white wine — or, to rescue a lost cause such as hard little wind-fall pears, bake them gently for 3 or 4 hours in red wine sweetened with sugar.

• To restore the flagging spirits of the cook in times of crisis.

The most important point of all about wine in cooking, is that the wine itself should be properly cooked — in other words, the alcohol must be driven off by boiling, leaving only its fragrance behind in the pan. If wine is added to a casserole or meat sauce which will be cooked long and gently, the wine will volatilise of its own accord. But if it goes into the pan at the last moment, let it bubble fiercely to drive off the alcohol before adding any other ingredient. Or, if boiling would damage the contents of the pan, boil the wine separately until it is reduced by half before adding it.

Red wine has more uses in the kitchen than white. The very cheapest wines will serve most purposes (Spanish Burgundy, Algerian, Moroccan or an *ordinaire* from the Midi), though there are occasions when a special dish such as *coq au vin* is enhanced by the use of something slightly better.

Remarkably little wine does the job. There are few recipes which call for more than a glassful and it is always better to add too little rather than too much.

• FLAMES AND SPIRITS •

Although superfluous flames have become the hallmark of ostentatious restaurant eating, there is no flamboyance about flames in the kitchen. The proper purpose of sprinkling a piece of meat or chicken with warmed brandy and setting it alight, is to burn off the excess fat and to leave a residue which will combine with the juices in the pan to make a delicate sauce. It is not meant either to entertain or to impress those present. Anyhow, with most dishes that need to be *flambé*, the fireworks take place an hour or more before dinner.

• OIL AND FAT •

OLIVE OIL

Nothing distinguishes one tradition of cooking from another so tellingly as the oil or fat used for lubricating, frying and dressing food. Such is the penetrating, lingering nature of fats and oils, they imperceptibly characterise everything edible they touch.

Of all the nutritive oils, the juice of the olive, which is recommended throughout this book as a cooking medium, is the most digestible; also it is the most satisfying.

Olive oil varies in quality and price nearly as much as wine, and for much the same reasons. Both have the same enemies (late spring frosts and summer hail) and the same fluctuating yield.

The virgin oil from Provence — which is the product of the first pressing from olives hand-picked at that crucial moment between ripeness and maturity — is generally agreed (except perhaps by the olive farmers of Lucca in Italy) to be the finest of all. It is the olive oil equivalent of a great vintage *mise en bouteille au château*. And it's almost as rare and should be preserved for a special occasion and special dressing.

The cheapest olive oil comes from Spain, but for many tastes Spanish oil is rather too fruity as the Spanish emphasis is on quantity rather than quality.

For general purposes and for people who like to recognise the presence of the olive but do not wish to be overwhelmed by it, an Italian oil (such as San Remo) selling at about 39s. a gallon is the ideal compromise for both palate and pocket.

Sasso is a highly refined oil which would suit all those who desire the nutritional and other qualities of olive oil, but who dislike its taste. But it is more expensive than the unprocessed product. Sasso seldom sells at less than 63s. a gallon. However, a bland blend from Boots costs only 23s. for just under half a gallon.

It always pays to buy olive oil in quantity. For example, a gallon bought in 24 one-third of a pint bottles works out at 90s. a gallon. The same amount bought in one can costs only 59s. a gallon.

A good economic compromise is to buy a gallon can of medium-priced oil for all ordinary cooking purposes while treasuring a costly bottle of the finest oil for *vinaigrette* dressings.

BUTTER

If you like the taste of butter, and I do, it has no substitute — especially in the kitchen. Just as the telephone brings out subtle inflexions of a voice, so heat widens the taste gap between real butter and its imitators.

Remember :

• Salted butter keeps better than fresh, but in cooking, allowance must always be made for its saltiness.

• Butter burns easily, but if you want the flavour of butter, without this disadvantage, add a spoonful of bland olive oil to the melted butter before adding the other ingredients.

• CREAM •

SOUR CREAM

Sour cream is one of the most versatile commodities in the kitchen.

Unfortunately cream that has soured naturally is not easily come by these days because pasteurisation kills the natural souring agent. Instead of going pleasantly sour, cream sold by most of the big dairies first separates, then grows whiskers and finally putrefies.

Though it is still possible to buy unpasteurised cream direct from a farm (or, for Londoners, from Wholefood in Baker Street), for all kitchen purposes cream which has been artificially soured with a special culture, or fresh cream soured with lemon juice serves just as well.

Fresh and slightly edgy in character, sour cream blends as happily with sweet foods as it does with savoury, adding a touch of astringency to fruit and puddings and a blandness that never cloys to soups and sauces.

Try sour cream:

> with fruit soft or stewed for breakfast
> stirred into a spaghetti sauce or a *ragoût*
> poured over blinis or pickled herring
> added to fresh vegetable soups
> as a salad dressing (5 parts sour cream to 1 of lemon juice or wine vinegar, seasoned with fresh herbs, salt and freshly ground pepper)
> as an alternative to fresh cream when mixing horseradish sauce
> as a dressing for green beans, or stirred into peas.

See also : Mushroom filling for pancakes (*see p. 81*)
Paprika plaice and sour cream (*see pp. 113–14*)
Honey cake (*see p. 241*)
Fried Aubergine (*see p. 163*)
Sour cream and mushroom sauce (*see p. 191*)
Sour cream and cucumber sauce (*see p. 191*)

FRESH CREAM

Single cream is less likely to separate when added to hot dishes than double cream. Thick Jersey cream is a table luxury and quite unsuitable for most cooking purposes.

When adding cream at the last minute to a hot soup or sauce, boil it briefly in a separate pan to thicken it slightly, but do not reboil the dish after it has been added. If there are acid elements (wine, citrus juice, tomatoes) in the dish, work a little flour into the cream before it is boiled ($\frac{3}{4}$ tsp to 1 gill).

· GARLIC ·

This is the most aggressive member of the lily family of which, as a nation, we are still strangely shy. Though it makes many appearances in this book, garlic is quite unobtrusive in most of the recipes requiring it, being just part of the subtle build-up of flavours. For instance, it shows off the savour of cooked cheese without really drawing attention to itself; while the flavour of mushrooms and, to a lesser extent tomatoes, seems to expand under the influence of garlic.

Try drawing a cut clove across the dish that is to hold a cheese *soufflé*, the pastry crust beneath a cheese flan or the toast that supports a Welsh rarebit; or spread a little garlic butter over the gills of a large flat mushroom before putting it under the grill. Like good tailoring which is said to be more easily recognised than described, the garlic in these dishes should be sensed rather than tasted.

···*···

For many people a salad unkissed by garlic is a sorry thing. But if you are shy of the bulb, or suspect your guests may be, you can give the green leaves an aroma of garlic without actually mixing it into the dressing.

Squeeze the juice of a small clove on to a square inch of bread that has been soaked with a few drops of olive oil. Put this *chapon* (as the morsel is called) at the bottom of the bowl and pile the salad on top of it. Within half an hour the *chapon* will have done its work and can be discarded (although in some households it is a tit-bit to be competed for when the salad is finished).

···*···

Only a few dishes in this book are unashamedly garlicky : *aioli*, *bourride*, *salsa verde*, *pesto sauce*, *garlic butter*, *spaghetti* dressed *al aglio e olio*. They are included for the true *aficionados* of the bulb — a growing fan club.

Poulet à l'ail from Béarn, which contains ½ lb of garlic, is not included in the list above because the bulb is blanched before cooking, which makes it almost innocuous.

• HERBS •

To the discerning cook, fresh herbs are indispensable — and within reach of anyone prepared to sacrifice three square feet of window sill or balcony to accommodate three or four flower pots or a window box, or a small corner of a kitchen garden.

The habit of using fresh herbs in the kitchen is as British as the Belvoir Hunt. In the days when spices were rich men's luxuries,

ordinary Englishmen had to rely on herbs to flavour their food. Writing in the 16th century, Thomas Tusser, the A. G. Street of his day, selected from his list of some 300 aromatic garden herbs a mere 42 which he reckoned no cook could possibly do without.

Modern Londoners, and other city dwellers, are forced to be even more selective.

In the most literal sense, window box cultivation is suitable only for the down-to-earth herbs such as thyme, tarragon, marjoram and chives. And, of course, parsley and mint if space can be spared for the greengrocer herbs.

Plants like fennel and sorrel, which have a tendency to shoot up and run to seed in a window box, are best left to gardeners with gardens.

Rosemary thrives in city conditions, chervil tolerates them, winter savory flourishes through the summer (though in my box it has never lived up to its name) and bay trees behave like any other evergreen, provided their roots are watered in the summer. (If you do not feel like investing some 3 gns in a decorative tree, ask your nursery for a small culinary bay costing about 10s.)

Basil, the most fragrant and temperamental of them all, can be raised when Ceres is feeling benevolent and the weather well behaved (a coincidence of events which occurs in my corner of London approximately every other year). But for the sake of having these sweet, spicy leaves with their intoxicating scent on my window sill, I would go through the drill of planting and cherishing basil even if I was only rewarded once in seven years.

City husbandry, which is my only experience of gardening, has rules all of its own, and the first and most important rule concerns the soil.

Don't be tempted to beg a little bit of your neighbour's earth, or to slink off, shovel and sack in hand, to the nearest park or vacant lot. Acid soil makes a most inhospitable bed for young plants.

Use a compost composed of medium loam mixed with leaf mould, sharp Bedford sand and a compound fertiliser. Enough to fill an ordinary box should cost 6s.-7s. from any nursery.

Herbs need a deep, well-drained box, a reasonably sunny situation and a complete change of soil every fourth year. In the intervening years replace the top third of the soil with fresh compost, working over the new and the old until they are perfectly blended.

Basil and chervil, which may have to be grown from seed as not all nurseries keep the plants, get a better and quicker start in life if they are sown in sterilised soil in a small seed box. The cost of these extras can be counted in pennies, the results in the number of healthy young plants.

Fill the box with John Innes sterilised compost No. 1 (obtainable from your local garden shop) and scatter the seeds very thinly over the surface. As soon as the seedlings are stout enough to handle, transfer a few of the most robust to small pots for a few weeks before settling them finally into the window box with the rest of the herbs.

If you cannot buy plants locally, write to a seed specialist such as Sutton & Sons Ltd. of Reading; Unwins, of Histon, Cambs; Dobbie and Co., of Edinburgh, 7; or Thompson and Morgan, of Ipswich.

However, the wise window-box gardener will always opt for plants rather than seeds if he has the choice. Luckily most herbs can be bought in the small plant stage, ready to go straight into the prepared box — April and May is the time to plant most herbs, though basil shouldn't go in till June — one frost and it's finished.

Most nurseries of any standing carry a modest range of herb plants, and there are a few specialist firms, such as the Herb Farm at Seal, Kent, which could aspire to plant up a herb patch for Thomas Tusser.

Evidence of the growing popularity of fresh herbs is the increasing number of shops selling them fresh by the bunch. In London, bunches of fresh herbs in season can usually be found at :

Wholefood Ltd., 112 Baker Street, W.1.
Harrods food hall.
Selfridges' greengrocery department.
Roche, 14 Old Compton Street, W.1.
Blakes, 78 King's Road, S.W.3.

Fresh herbs by post. In season, Country Style, 18 Ship Street, Brighton, will send assorted packages at weekly or fortnightly intervals.

DRIED HERBS

Too often these look, and taste, just like sawdust. The oils which give each plant its characteristic taste are volatile and easily destroyed — unlike some of the dry-as-dust packets which appear to be everlasting.

A dried herb can never completely replace a fresh one and some, such as chervil, parsley and fennel leaves, are completely tasteless dried. With the exception of mint, they all acquire some degree of mustiness in the process of drying which makes them useless for any dish, such as an omelet *fines herbes*, where the fresh flavour is all-important. A sprinkling of dried herbs could ruin a salad, and parsley butter made with the dried leaves would be just a waste of good butter.

However, there are some which survive the process of dehydration better than others, notably thyme, tarragon, origano, rosemary and basil.

Dried herbs for winter use should be chosen with discrimination and bought in small quantities. They keep their flavour longest if they are stored in tiny, airtight tins or glass jars. The Culpeper shops are good suppliers. The ¼ oz drums prepared by Aroma Culinary Products (the range obtainable is wide) may seem unnecessarily expensive, but by buying in small quantities it is possible to use the herbs up before they lose their savour. Perhaps the most effective dried herbs I know are those put up by the Chiltern Herb Farm, though I regret they have seen fit to increase the size of the drums. It just means there are more herbs to throw away at the end of the winter. Because, however effective dried herbs are when first bought, these delicate fragments hold their aroma for only a few months, so do not expect them to survive for more than one season.

If you are lucky enough to come across a shop selling little, faggot-like bunches of herbs dried on the stem, snap them up. Herbs dried in this manner retain much more of their pungency.

IMPORTANT NOTE. Use all herbs with discretion, and rosemary with extreme caution. Too much of any herb can ruin any dish, and too much rosemary can ruin the whole meal by irritating the palate. Rosemary, like cloves, is best pulverised before use. Three to four spikes are enough to perfume 3 lb of lamb or pork. Alternatively, use a sprig of rosemary as a brush for oiling meat or fish.

Warning

• *Laurier* in French recipe books should be translated as bay, **not** laurel.

• The sweet scented geranium referred to in chapters 3 and 12 must not be confused with the garden and window box variety. The scented geranium (*pelargonium capitatum*) is grown for its leaves, which give off a sweet spicy scent, a blend of lemon, rose and balsam. They are oak leaf shaped and hairy. Its tiny flowers are pink.

• SEASONINGS •

SALT — sea salt or *gros sel*. This is an unadulterated salt, sold in crystalline (not free running) form. Its clean, stimulating flavour quickly becomes an addiction (a healthy one) to anyone who uses it for more than a week. Cost — about 6d. a lb. Buy 2 or 3 lb when you can, and keep it in a confectioner's sweet jar. Sold in Soho, also by Harrods, and Jacksons of Piccadilly.

PEPPER. Ground pepper, like ground coffee, loses its aroma very quickly. Buy black (the most aromatic) and white (the hottest) peppercorns by the ounce, and keep two grinders, one dark and one pale to dispense them. There are many, many more uses for black pepper than there are for white. An ounce of white pepper lasts me more than a year.

MUSTARD. Keep English mustard powder (if you like English mustard) and French mustard from Dijon or Bordeaux (if you like French mustard), but avoid English imitations of French mustard. The subtleties of French mustard are a small study on their own, the French showing their characteristic ingenuity in both the liquid used for mixing and the herbs and spices employed for flavouring. *Maille* and *Grey Poupon* are the two most famous names in French mustard. When adding mustard to a sauce, always use one that is pale yellow.

VINEGAR. Malt vinegar is excellent for rinsing hair and cleansing sponges, but it is not suitable for cooking or for dressing salads. Vinegar for kitchen use should be made from wine, or, which is cheaper, cider. If you choose wine vinegar, keep one bottle each of white and red and one bottle that is flavoured with tarragon. If you grow tarragon, this is easily prepared at home. Just stuff a bottle with fresh tarragon leaves on the stem, and fill it up with good white wine vinegar.

··· * ···

VANILLA SUGAR. Sugar is a seasoning as well as a sweetener (see what it does to peas, carrots and tomato soup). Vanilla sugar (which is sold in packets at the extortionate price of 9d. an ounce — which my mathematics tells me works out at 12s. a pound) is easily made at home. Just keep 2 vanilla pods in a 1 lb jam jar filled with castor sugar and top it up every time some is used. Many of the recipes in this book suggest vanilla sugar, as it has a gentle aroma which emphasises the subtleties of many fruits, particularly apricots.

· SPICES, SEEDS AND STIGMA ·

As with peppercorns, these should be bought whole and ground or crushed as they are needed. Use either a pepper mill or a pestle and mortar (the wooden ones are most convenient for this), or, as a last resort if you have neither, wrap them in a piece of cotton and crush them with a mallet as you would ice.

For spicing mulls and punches the whole spices can be dropped straight into the liquid, provided it is going to be strained before serving. For fruits and pickles, constrain the spices in a muslin bag which can be removed.

A FEW OF THE MOST USEFUL SPICES

ALLSPICE. (alternatively called "Jamaica pepper" or "pimento" not to be confused with paprika which is made from dried sweet capsicum peppers). Useful in *pâtés*, sausages, soups, mulls and marinades for meat.

CARAWAY. For gulyas; cabbage — red or white; adds an interesting note to soft cheesy mixtures, both hot, like Welsh rarebit, or cold, such as potted cheese and cheese spread.

CINNAMON (made from the bark of the cinnamon tree). Keep cinnamon in both its forms, whole and powdered. It is difficult to

S.T.C.B.—C

pulverise cinnamon effectively at home. It's the top spice as far as mulls are concerned, but don't overdo it — this bark has a powerful bite. Add 1 inch of cinnamon to the water when cooking bacon, ham or boiled chicken and just a touch mixed with sugar on milk puddings, junkets and plain boiled custard.

CORIANDER. For anything that gets a boost from a whisper of orange — pork (roast or chops), gravy for a duck, sole — just a touch in the sauce, mashed potatoes, apples baked or stewed. Splendid for seasoning stuffings for poultry, *pâté* and homemade sausages. And for an experiment try :

- mixing coriander with garlic and caraway as a seasoning for fish
- mixing 1 dozen seeds with 1 lb of coffee beans before putting them into the grinder.

JUNIPER (the berry that gives gin its name and flavour). Use sparsely as a seasoning for game, pork and veal; add to marinades for meat; try sprinkling a little over potatoes as they cook, whole or sliced in butter.

MACE (the outer casing of the nutmeg). Use a blade or two for flavouring the poaching liquor for silverside of beef, chicken, ham or fish; it is an important flavouring for milk to be used in white sauces and a good seasoning for oyster soups and stews, liver *pâté*, mashed potatoes and mushrooms.

NUTMEG. A dead ringer with anything cheesy, or anything made with spinach (soups, *soufflés*, *gnocchi*, etc.).

PAPRIKA. The sweet pepper made from dried, sweet capsicums. Buy the mild Hungarian paprika and it will never burn your tongue. However, it burns very easily itself so remove the pan from the heat before sprinkling it directly on to onions, meat, etc. Paprika is indispensable for gulyas. It is also useful for seasoning and colouring sauces for fish, veal and some vegetables. Try sprinkling a little at the last moment on a panful of fried aubergines or courgettes.

SAFFRON. Prized for its colour as well as the touch of bitterness it imparts to anything it touches. One of its traditional roles in British cooking was the name part in Cornish saffron buns. They still have yellow buns in the Duchy, but their yellowness owes nothing to saffron which is too precious to fritter away on something as common as buns ! Saffron is made from the stigma of autumn crocuses and it takes something like a quarter of a million blooms to produce 1 lb of saffron. Commercially it is used these days for colouring certain liqueurs such as yellow chartreuse. Its main contribution in the kitchen is enriching the tint and taste of *risottos* and *paellas*, fish soups and sauces.

Because it is so expensive, saffron is often simulated, but only in its powdered form. Buy the little filaments and you will know you're getting the real thing. Boots sell the cheapest saffron I've ever found.

To use: Crush the stigmas and steep them for 15-20 minutes in warm water or stock or a little of the liquid to which it will be added.

• SHOPPING FOR STORES •

The store cupboard is the emergency fuel reserve and, wisely equipped, it can simplify the whole business of housekeeping as well as arming one against every contingency from sudden guests to sudden laziness, from Bank Holidays to bankruptcy (temporary, of course).

The list that follows omits all such conventional stores as sugar and flour, which go on to the grocery list automatically.

DRIED FOODS

RICE : Rice is the most hospitable cereal. It combines amicably with almost all meats and vegetables and, if there is rice in the cupboard, it is always possible to rustle up a meal of sorts. It is useful to keep both Patna (the long grain rice for *pilaffs* and for boiling) and Po Valley rice (the round, absorbent grains for *risottos*).

PASTA : Keep some *lasagne* in the cupboard (the wide, ribbon *pasta*, either green or white, for baking with meat or tomato sauce), as well as long spaghetti and skeins of egg noodles which are a convenient alternative to potatoes as an accompaniment to casseroles and *ragoûts*.

PULSES : Brown lentils, split peas and haricot beans are invaluable for soups, casseroles or *purées* to accompany roast meat and game. Lentil *purée* marries most effectively with partridge, venison, hare and pork. It is also good dressed with *vinaigrette*. Lentils, incidentally, don't have to be soaked; on the other hand, two or three hours in water will cut down cooking time. If you are really rushed, they can be pulverised in the blender. Haricot beans, a natural with lamb or mutton, and essential to *cassoulet*, need 8-12 hours' soaking; so do chick peas.

DRIED CHESTNUTS : For adding luxury and body to stews.

DRIED CÊPES : For soups and sauces.

DRIED APRICOTS : For flans, *soufflés*, sweet sauces.

NUTS : Shelled, whole, unblanched almonds and just a few blanched, shredded almonds (for adding to *sauté* potatoes or for trout with almonds), and pine nuts (*pinoli*). These are small, milk coloured fruit of the pine tree with a warm, slightly resinated taste. An important constituent of *pesto sauce* (*see p. 192*), a pleasant surprise in rice salads; good fried in butter and stirred into cabbage.

TINNED FOOD

Because foodstuffs vary so enormously in their reaction to the canning process, zealous discrimination is essential when shopping for tins. Delicate berries, subtle vegetables and certain foods high in starch lose their essential characteristics during the prolonged sterilisation at 240 deg. F. and upwards to which all tins have to be submitted in order to destroy the bacteria. With these, the metamorphosis is so marked that the name and the picture on the tin bear almost no relation to the texture and taste of its contents. Strawberries, spaghetti, rice, artichoke hearts, peas (except for imported *petites pois à la Française*), aubergines, asparagus, round tomatoes, chicory, green beans, porridge and potatoes all come into this category.

Howevei, many foods suffer canning gladly and emerge almost unscathed from the experience. These are the foods to buy for the cupboard, and the list (including brand names) that follows offers just a few suggestions based on personal experience.

• Good quality meat or game soups such as Baxters.

• Clear *consommé* for emergency sauces or for serving cold in the summer, topped with a little floating cream and a sprinkling of chives or spring onion tops.

• A selection of seafood for *risottos*, *pilaffs*, *hors d'oeuvres* and salads — anchovies and sardines from Portugal, Norwegian shrimps, fillets of smoked eel, French tuna fish, Princess crabmeat, mussels and clams. Also tinned salmon from Canada for kedgeree, fish loaf, *soufflés* and *vol au vent*.

• Soft roes.

• A tinned, boneless ham (Plumrose do a beautiful one).

- *Prosciutto* — raw smoked ham from Italy. Expensive, but the perfect grand starter for unexpected guests. Serve with a lump of butter or with melon.

- Sweet pickled cucumbers and small pickled beets.

- *Pimentos:* for mixed salads (with olives and tomatoes); for adding to the juices from a roast or grill; added to omelettes (1 tblsp for every 2 eggs); as part of an *hors d'oeuvre* with smoked eel, hard-boiled egg and onion rings.

- Black olives.

- Tins of button mushrooms and, if possible, *chanterelles* (ochre coloured trumpets with a rich autumnal flavour) for adding to cream sauces with chicken or veal.

- Leaf spinach for *soufflés, gnocchi* and soups, as well as on its own as a vegetable.

- Corn kernels (Green Giant Niblets or Mexicorn) to combine with rice or peas, as a bed for a poached egg, combined with mayonnaise or as a vegetable in its own right.

- Italian tomatoes for all tomato sauces and casseroles demanding tomatoes. Cirio are one of the best brands.

- Tins of *bolognese* and tomato sauces for *pasta*. The GO brand is good.

- Faugier's fresh, unsweetened chestnut *purée* : for soups, stuffings; *purées* to accompany pork, turkey or game; or as a pudding sweetened and flavoured with rum or brandy or grated orange.

- Japanese white peaches, the peachiest peaches in tins.

- Ritz meringues. These are the real thing, made with egg whites and sugar, and they stay fresh almost indefinitely.

TUBES

These are sensible containers for certain perishables which are damaged by contact with the air :

Mayonnaise (Reymersholms).

Horseradish (for dilution with cream).

French mustard from France for picnics and emergencies.

Caviar (for entertainment only).

Chestnut (for decoration).

• A FEW DEFINITIONS •

AL DENTE : Food cooked to the point where it still offers some slight resistance to the teeth. Usually applied to *pasta*.

BAIN MARIE : A shallow pan of hot water, big enough to hold other pots and pans. It is used for cooking gently (as with custards and *pâtés*) and for keeping things such as sauces warm.

BEURRE MANIÉ : A blend of flour and butter in the proportion of 3 : 4. Used for thickening liquids at the last minute. Add it in small raisin-sized lumps, stir till dissolved and do not reboil the liquid. ¾ oz flour and 1 oz butter should thicken 1 pint of liquid.

BLANCHING : Immersion in boiling water for ɪ

• removing skins as with almonds, tomatoes, etc.

• whitening as with brains, calf's feet, vegetables

• to remove saltiness as with bacon.

BOUILLI : The beef cooked in a *pot-au-feu.*

BOUQUET GARNI : A bunch of herbs used for flavouring sauces, soups and casseroles. Usually consisting of thyme, bay and parsley, it can also include celery, marjoram and other herbs.

CHAPON : A crust of bread soaked in oil and impregnated with garlic. Used for scenting salads (*see p. 28*).

COCOTTE :

• a deep lidded casserole made of metal or earthenware

• small porcelain dishes with handles for the baking or steaming of eggs.

COURT-BOUILLON : A mixture of wine, water and aromatics, used for the poaching of fish.

CROÛTONS : Pieces of bread cut in small squares or large triangles, and fried in butter. The squares are sprinkled in soup, the triangles accompany dishes with rich, creamy sauces.

CRUDITIES : Raw vegetables.

GRATIN (AU) : A dish sprinkled with breadcrumbs and butter or oil and browned under the grill or in the oven. It has nothing to do with cheese.

GRATIN DISH : A shallow, oval dish made of earthenware or metal, suitable for putting under the grill.

LARDING : Threading small strips of fat into lean meat before roasting. Sometimes bits of fat are seasoned with herbs or spices to enhance the flavour of the meat. Though usually threaded in with a larding needle, the bits of fat can be inserted into a slit made with a sharp pointed knife.

MACERATE : Similar to marinate, but applied to the soaking of fruit, usually in wine or spirit.

MARINADE : An aromatic bath of wine, spices, herbs and vegetables for uncooked meat and fish. The object of marinating is to :

- tenderise tough meat
- moisten dry flesh as with hare, for instance; when moistening is the main object of the exercise, olive oil is usually added to the marinade
- impart the appropriate aroma to meat or fish.

Meat can be left in a marinade for anything from 1 hour to several days, depending on the weather, the animal and the state of its flesh.

Fish is rarely marinaded for more than 1 hour. The mixture usually contains oil, and either white wine or lemon juice.

POACH : To cook in liquid without boiling. Fish, as well as eggs, are frequently poached.

RAGOÛT : A more appetising word for stew.

REDUCTION : Concentrating a liquid by fast boiling.

ROUX : A cooked liaison of flour and butter (using marginally more butter than flour). Used as the basis of sauces, *soufflés*, gravies, etc. A white *roux* is the colour of Devonshire cream; a medium *roux* is sherry coloured; and a brown *roux* the shade of a hazelnut.

SAUTÉ : Literally "to jump". In fact, it means to cook very slowly in shallow fat such as butter or oil in an open pan. The jumping refers to the need to keep the pan moving so that the food won't stick. It is both a preliminary treatment — mushrooms, for instance, are usually *sautéed* before being added to a casserole — and a way of cooking as for potatoes and young vegetables.

SEETHE : Old English for gentle stewing.

TERRINE : Used to describe both the dish in which a *pâté* is cooked and the *pâté* itself.

ZEST : The oily outer skin of citrus fruit containing both the colour and the flavour.

• WEIGHING AND MEASURING •

It is sometimes implied, rather foolishly, that weighing ingredients is the hallmark of an L-plate cook, that the clever woman cooks by intuition, throwing in a little bit of that and a pinch of this. This attitude probably derives from the general vagueness of French cookery books which are inclined to talk of *un petit verre* without specifying just how *petit* it should be In fact, such imprecision occurs mainly in recipes for soups, casseroles, *pâtés* and similar dishes where there is plenty of scope for imagination and only few immutable rules. In recipes where precision really matters, such as *pâtisserie*, French cookery books give accurate quantities.

But whatever you are making, if the formula is strange it is wise to measure the ingredients carefully the first time you try the recipe. To understand a rule well enough to break it, you must submit to it at least once. And for cakes, preserves, pastry, all sugar cooking and some sauces, accurate measuring is essential.

This is not meant to be an argument against improvisation, which is the soul of cookery, only a cautionary word against anarchy in the kitchen.

SCALES

Balances that work with weights are more dependable, robust and versatile than spring balance scales; also small quantities, where accuracy is essential, can be more precisely gauged with the aid of weights. There is the added advantage that when working with foreign recipe books you can use gram weights instead of involving yourself in the higher mathematics of converting ounces and pounds into grams and kilograms. French weights are sold by the sort of kitchen shops which cater for professionals as well as amateurs.

In case you have forgotten —

1 kilo	=	**2 lb 3 oz approximately**
100 grams	=	**3½ oz approximately**
1 litre	=	**1¾ pints approximately**
1 decilitre	=	**6 tablespoons approximately.**

MEASURING CUP

This is a must in any kitchen. Be sure you get the British Standard measuring cup made of Pyrex Glass. It holds ½ pint (10 fluid ounces), and should not be confused with the American cup measure which is based on the 16 oz American pint. This disparity between British and American pints (and measuring cups) must always be remembered when working from American cookery books.

1 British cup	=	**1¼ American cups**
1 British pint	=	**1¼ American pints**
1 British gill	=	**1¼ American gills**

Happily,

1 pound (lb) = **1 American pound (lb)**

But

tablespoons are marginally larger than American tablespoons.

MEASURING SPOONS

A set of 5 linked spoons, ranging from ¼ tsp to 1 tblsp which, unlike your tableware, are made to the British Standard. Spoons are the only way of measuring unweighable quantities (such as $\frac{1}{16}$ of an ounce) and the quickest way of counting out the small quantities.

Liquid

1 tablespoon	=	4 teaspoons	= ½ fluid ounce
1 teaspoon	=	⅛ fluid ounce	

Floury substances such as ground spice

1 teaspoon	=	⅛ ounce
½ teaspoon	=	1⁄16 ounce

A FEW USEFUL WEIGHTS AND MEASURES

LIQUID

1 British Standard cup	= ½ pint or 10 oz or 2 gills
1 tablespoon or 2 dessertspoons or 4 teaspoons	= ½ fluid ounce
1 gill	= 5 fluid ounces
3¼ cups	= 1 litre

SOLIDS

SUGAR (castor)

1 British Standard cup	= 8¼ ounces
1 tablespoon	= ½ ounce

FLOUR

1 British Standard cup	= 6½ ounces
1 tablespoon	= ½ ounce

TREACLE AND HONEY

1 British Standard cup	= 14 ounces
1 tablespoon	= ¾ ounce

RICE

1 British Standard cup	= 9 ounces
1 tablespoon	= ¾ ounce

BUTTER

1 tablespoon	= ¾ ounce

Cup measurement omitted here because I do not believe butter can be measured accurately in cups.

FRESHLY GRATED PARMESAN CHEESE

1 tablespoon	= ¼ ounce

SULTANAS AND CURRANTS

1 British Standard cup	= 7 ounces

GELATINE

Recipes in this book requiring gelatine all specify leaf gelatine, as I find it rather simpler to use than the granulated type. But whichever kind you buy, always get the best.

6-7 thin sheets gelatine   =   2 level tablespoons granulated gelatine   =   1 ounce

Allow approximately

½ oz gelatine to set 1 pint liquid
¼ oz gelatine to set 1 pint creamy mixture such as a mousse.

THERMOMETER

Certain temperatures in cooking are too critical to be left to guesswork. An excess of heat or not enough heat can ruin food, wasting time and ingredients. A kitchen thermometer costing less than £1 does away with this element of chance and can earn its keep over and over again.

Choose one made of stainless steel, with a non-conducting handle and a sliding clip for fixing the thermometer on to a saucepan.

Points on the Fahrenheit scale worth remembering

100 deg.	Beyond this the coagulating power of rennet diminishes.
162 deg.	The point at which germs cannot survive in milk. Most milk is pasteurised these days, but this is a figure worth remembering if you are on holiday and can only get unpasteurised milk.
150 to 160 deg.	For the water sterilisation of bottled fruit.
205 deg.	Curdling point of custard. Simmering point of water and stock. Not to be confused with boiling point which is 7 deg. higher.
220 deg.	Setting point of jam
320 deg.	Minimum temperature for deep frying — below that the fat will penetrate the food — for many foods the temperature is higher.
392 deg.	Burning point of lard ⎫ *Never allow fats*
	⎬ *to exceed their*
554 deg.	Burning point of olive oil ⎭ *burning point.*

OVEN TEMPERATURE CHART

Description	Regulo Setting	Degrees Fahrenheit
Very cool	$\frac{1}{4}$ $\frac{1}{2}$ 1	240 265 290
Cool	2	310
Warm	3	335
Moderate	4	355
Fairly hot	5 6	380 400
Hot	7	425
Very hot	8 9	445 470

FOOD

-4-

SOUPS
AND STARTERS

• SOUPS •

I have only one complaint against tinned soup and it concerns not the contents of the tins (which are generally good, and often excellent), but the depressing effect they have had on the general status of soup as a course. Because it mostly comes out of a tin, soup these days, with the possible exception of *crême vichyssoise*, has been de-classed in this country to the point where it is rarely served on elegant occasions.

Yet of all possible ways of starting a meal, grand or otherwise, it is hard to better a delicate soup made with fresh vegetables, home made stock or milk, cream and herbs.

✳ *Lettuce Soup* ✳

Lettuce is the basis of one of the lightest of summer soups.

For 4 people

> 2 lettuces (or the outside leaves of 4)
> 4 spring onions
> 2 oz butter
> ½ pint each milk and water (or 1 pint good chicken stock)
> 1 gill cream
> salt, pepper and sugar

Shred lettuce finely, slice spring onions and add both to the butter, which has been set to melt in a heavy pan. After 10 minutes add the milk and water (brought to boiling point) and seasoning. Cover, cook for 15 minutes, then put the soup through a sieve or food mill (or blend in blender). Return to pan, add cream, adjust seasoning and serve.

* Spinach Soup *

> ½ lb spinach purée or 1 packet frozen spinach
> 2 tblsp butter
> 1 pint milk
> ¼ pint cream and seasoning

Cook 1½ lb well washed spinach in a covered pan without salt or water, then press it dry before putting it through a sieve or food mill.

Stir the warmed milk into the seasoned and buttered *purée*, let it simmer for 10-15 minutes and just before serving stir in the cream.

* Fresh Pea Soup *

> 1½ lb peas;
> 3-4 spring onions and 2 large lettuce leaves
> 1 gill each milk and thin cream
> salt, pepper and a pinch of sugar

Put the shelled peas into a heavy pan with the sliced lettuce leaves, spring onions, sugar and seasoning. Just cover the peas with water and simmer gently until they are soft. Strain off the liquid and put the vegetables through a sieve or a food mill.

Return the *purée* to the pan and dilute it with 1 gill of the liquid in which the peas were cooked, the milk and the cream. Heat gently. If the soup is to be served cold, make it slightly thinner by adding a little more liquid.

This soup is ideal for putting in Thermos flasks or as "one for the road" after a winter party.

* Tomato Broth *

> **12 oz tomatoes**
> **1½ pints stock**
> **1 small onion**
> **1 rasher of bacon**
> **bouquet garni with zest of orange, celery, bay, thyme, parsley**
> **¾ oz butter**
> **salt and pepper**
> **castor sugar**
> **2-3 tblsp dry sherry or 1 gill double cream**

Chop both onion and rasher of bacon and soften in ¾ oz butter with the *bouquet garni*. Roughly chop the tomatoes; add them and season with salt and pepper and a touch of castor sugar.

Leave the tomatoes to *sauté* for a few minutes before adding 1½ pints of stock (made from the bones of a bird or shin of beef), or one 15 oz tin of Baxter's pheasant or turkey *consommé* plus ¾ pint water.

Simmer for 30 minutes. Remove *bouquet garni* and put soup through a mill or sieve. Reheat; adjust seasoning; add 2-3 tblsp dry sherry and pour immediately into the heated flask.

If it is to be served at table, omit the sherry and stir in the gill of double cream.

* Beetroot Soup *

Also for picnics or after the party

> **1½ pints of stock**
> **1 onion, 1 leek, 1 carrot**
> **1 turnip, 2 sticks celery**
> **3 tomatoes**
> **bouquet garni of parsley and bayleaf**
> **1 large raw beetroot**
> **1 cooked beetroot**

THE STOCK : Make a rich stock with giblets of turkey or goose (plus some extra chicken necks if you can find them), a few bacon rinds, a pork rind, ¾ lb shin of beef and 2 pints of water.

S.T.C.B.—D

Chop the onion, leek, carrot, turnip, celery and tomatoes and add them to the stock with the *bouquet garni*. Simmer until the vegetables are tender.

Meanwhile peel and grate the large raw beetroot and cook it in just enough water to cover. When it is tender add the peeled and grated cooked beetroot and simmer both together for a few minutes before pressing the mixture through a food mill or sieve. Return the beetroot *purée* to the pan, strain the stock and add it to the *purée*. Stir in 2 tsp concentrated tomato *purée* (or a small tin of sieved Italian tomatoes), salt and pepper. Bring almost, but not quite, to the boil and turn into a heated flask.

If the soup is to be eaten at the table, serve sour cream or yoghourt separately.

* Oyster Soup *

   12 Portuguese oysters
   ½ gill milk
   1 tblsp each cream, butter and flour
   salt and pepper

THE STOCK

   1 cod's head; water to cover
   2 tblsp vinegar
   parsley, mace, salt and peppercorns

Simmer the ingredients for the stock together for 1½-2 hours.

Melt the butter, work in flour and stir till the mixture starts to honeycomb. Remove from heat and stir in gradually 1 pint of the fish stock diluted with the milk. Leave this to simmer for at least 15 minutes, stirring occasionally, then add the cream and the oyster liquid.

Just before serving throw in the oysters and adjust the seasoning. Once the oysters are added the soup must not boil, otherwise the oysters will be rubbery (boiling has the same effect on oysters as it does on eggs — the longer they boil, the harder they get). Serve the soup at once with *croûtons*.

For general notes on oysters see page 59.

* *Jerusalem Artichoke Soup* *

A remarkably interesting and popular soup

1 lb Jerusalem artichokes
1 oz butter
¾ pint each milk and chicken or veal stock, or 1½ pints milk
2 tblsp cream (unless all milk was used)
salt and pepper

Scrub and peel the artichokes (a tedious business) and put them roughly chopped into a heavy pan with the butter (already melted). Leave them to seethe gently till soft enough to press through a sieve or food mill.   Return the *purée* to the pan, work in liquid, seasoning, and simmer for 15 minutes.   Stir in the cream at the last minute. Serve with *croûtons*.

* *Croûtons* *

However good your soup, if it is creamy in texture it will taste all the better for the contrasting touch of a few crunchy *croûtons*. Cut a slice of bread (white or brown), remove the crust and spread both sides with butter, and fry till crisp both sides before cutting up into small squares.

• COLD SOUPS •

Lightness is the essential quality of cold soup, so it is important to remember that a hot liquid is liable to thicken when it cools.

If the preliminary cooking is done in fat, the soup should be most scrupulously skimmed before it is served.   Use butter rather than oil (it solidifies more readily), and use it sparingly.

Flour should never be used; a little thin cream will give a cold soup all it needs in the way of body and smoothness of texture.   The most attractive garnish is a scattering of finely chopped herbs and a few ice cubes — dropped in at the last moment.

The simplest and perhaps the pleasantest summer soup consists of a thin *purée* of seasonal vegetables, peas, beans, spinach, asparagus, tomatoes, etc., diluted with a mixture of milk and cream.

* *Gazpacho* *

More robust, but no less refreshing, is *Gazpacho*, from the plains
of Spain. A cross between a soup and a salad, it has a strong, clean
aste that is most appealing on a hot evening. There are any number
of versions. This one comes from Andalusia.

> $\frac{1}{4}$ cucumber
> $\frac{1}{2}$ red pepper
> 2 each tomatoes, black olives and cloves of garlic
> 1 tblsp wine vinegar
> 2 tblsp each olive oil and dried breadcrumbs
> black pepper, salt and cayenne pepper
> 1$\frac{1}{2}$ gills iced water
> chives or marjoram

Mince all the vegetables very finely. In a separate bowl work the
breadcrumbs into the creamed garlic before mixing in first the olive
oil (very slowly), and then the vinegar. Add this breadcrumb
mixture (which should be stiff enough to retain its shape) to the
purée of vegetables and season carefully with salt, pepper and a small
pinch of cayenne.

Put the soup to chill in a tightly-covered dish (to prevent the
garlic flavouring everything else in the fridge). Just before serving
thin the soup with the iced water, sprinkle with chopped chives or
marjoram and float a few cubes of ice on the surface. *Gazpacho* is
usually eaten with bread which is dipped into the soup.

* *Bortsch* *

From the opposite corner of Europe we get *Bortsch*, which, like
Gazpacho, has numerous variations. But two basic ingredients —
beetroot and sour cream — are common to them all.

> 1 pint good meat stock
> 2 onions
> 2 each raw beets and cooked beets (medium size)
> 1 oz butter
> salt, black pepper and celery salt
> bouquet garni
> 1 gill sour cream, or yoghourt

Shred one of the onions and both the raw beets and let them
sweat gently in the butter for 15-20 minutes. Add the stock, *bouquet*

garni and seasoning, and simmer slowly for about 1½ hours. Strain and leave to cool. Skim off the fat and reheat, adding half a cooked beet (finely shredded), the juice of both the onion and the rest of the beets. Check the seasoning.

To extract the juice from the onion — mince it and press the pulp through a sieve; from the beetroot — squeeze the minced pulp in a piece of butter muslin. Wash your hands immediately afterwards.

Do not let the soup boil. If it does it will turn brown, and its clear red colour is part of its beauty. Leave it to cool.

Serve chilled, with 1 tblsp of sour cream stirred in just before the soup is taken to the table. Hand the rest of the cream separately.

✳ *Iced Cucumber Soup* ✳

1 cucumber
½ onion cooked in ½ pint milk with a parsley sprig and 4 peppercorns
½ pint stock
2 tblsp olive oil and ½ gill cream, salt and pepper
chives and shredded cucumber — to garnish

Heat the oil in a heavy pan (oil is suitable in this case because of the milk) and *sauté* the peeled and sliced cucumber for 10 minutes. Meanwhile, slice the onion and let it simmer in the milk. When the cucumber is soft, add the stock and let the soup simmer in a covered pan for another 10 minutes before putting it through a food mill or sieve.

Stir in the strained milk and adjust the seasoning. Let it cool before putting it in the fridge to chill. Before serving stir in ½ gill of thin cream and garnish with chopped chives and shredded cucumber.

✳ *Iced Orange and Tomato Soup* ✳

2 lb ripe tomatoes or 2 tins Italian tomatoes
2 spring onions
1 orange
1 tsp each white wine vinegar and white sugar
1 glass white wine
1 gill cream
salt, pepper and tarragon (if available)

Sieve the tomatoes and simmer the *purée* for 10-15 minutes with the chopped spring onion, tarragon and a small thin slice of orange zest. Strain the *purée* and add all the other ingredients, leaving the cream until last. Chill thoroughly and garnish with a little chopped tarragon.

* *Yoghourt Soup* *

This is a cross between an iced soup and an hors d'oeuvre. *It makes a cool starter to summer meals, and involves no cooking.*

For 6 people
>   4 gills yoghourt
>   1 gill double cream
>   1 large cucumber
>   ½ pint shrimps or prawns
>   a handful of very fresh mint
>   salt, pepper and garlic

Mix yoghourt and cream together, stirring vigorously. Peel the cucumber (which should be as cold as possible) and shred it into slivers the size of a toothpick before stirring it into the yoghourt.

Add the peeled shrimps or prawns, three-quarters of the mint (finely chopped) and salt and pepper. Leave this in the fridge to chill.

Just before serving, work in half a creamed clove of garlic, adjust seasoning if necessary and sprinkle the remaining mint over the surface.

• STARTERS •

The prime function of the starter to an English meal is to stimulate the appetite, not satiate it, to play the meal in not play it down. *Pasta*, which was designed to dull the craving for costlier foods, is not a suitable starter for adults in the affluent society. It can, on the other hand, make a magnificent main course, as do so many of the provocatively savoury dishes loosely classified as *hors d'oeuvres* which are so substantial or rich, or both, that they detract from the main dish instead of leading up to it. To put temptation temporarily out of sight, I have put all these in the next chapter.

The perfect *hors d'oeuvre* is fresh looking, piquant and small, such as the yoghourt and shrimp iced soup given above, crisp vegetables dressed with *vinaigrette* (*see p. 181*), or a selection of *crudities* accompanied by a few fresh shrimps, a little smoked meat or a slice or two of continental sausage.

A few *crudities* (finely grated or chopped raw vegetables) to choose from, there are many others :

> **Plain white celery,** and the mildest of true French mustard to go with it.
>
> **Radishes,** red.
>
> **Radishes,** black, grated with a coarse grater and dressed with cream, black pepper and lemon juice.
>
> **Cucumber,** thinly sliced, and seasoned with salt, pepper, oil and vinegar.
>
> **Black olives.**
>
> **Baby carrots,** pared, grated with the finest rasp you have, dressed with oil, lemon juice, chopped spring onions, salt and a touch of sugar.
>
> **Celeriac,** peeled, grated into acidulated water, drained and blanched in boiling water for 2 minutes, drained again and folded into a mustardy mayonnaise.

One of the freshest and most appealing of first courses is a vegetable cooked *al dente*, dressed with a *vinaigrette* sauce.

Candidates for this treatment are legion — from such obvious and luxurious ones as asparagus and artichokes via slender leeks, baby sprouts and buttom mushrooms (these are served raw) to one of the best of all vehicles for *vinaigrette*, the flowerets of cauliflower.

✳ *Cauliflower* ✳

Divide a firm cauliflower into its individual flowerets. Wash well. Plunge it for 2-4 minutes into boiling water seasoned with salt and lemon juice. The cauliflower must retain its crispness. When it is drained and cold, turn it in a thick *vinaigrette* sauce :

> **3 tsp Dijon mustard**
> **½ gill olive oil**
> **1 dessertsp white wine vinegar**
> **salt, pepper, 2 shallots**

Put the mustard into a bowl and thin it slightly with ½ tsp of the vinegar. Then start adding the oil, drop by drop, and then in a steady trickle, adding the vinegar every so often until it is used up.

When all the oil has been absorbed, add the seasoning and chopped shallot. Its consistency should be that of a thin mayonnaise.

Make sure every floweret of the cauliflower is well coated with sauce, garnish with parsley and leave it to get well chilled.

* Leeks *

Choose leeks no thicker than a plump piece of asparagus (about 7 to the lb). After cleaning them (*see p. 171*) plunge them into boiling water seasoned with salt and lemon juice, cover and simmer for 6-10 minutes, according to their size. Drain thoroughly, pressing the water out with a little gentle pressure of the fingers. Leave them green end downwards to cool in a colander. When quite cold serve with a *vinaigrette* sauce made with :

> 1 tsp each Dijon mustard, lemon juice and parsley
> 1 dessertsp each white wine vinegar and chopped shallot
> 4 tblsp olive oil
> salt and pepper

Spring onions can be served in the same way, but they need only 60-80 seconds in boiling water.

Sprouts. Choose sprouts no larger than a muscat grape. Remove stalk, blanch for 2-4 minutes in boiling salted water and serve with the *vinaigrette* given above for leeks.

Mushrooms. Only small, white mushrooms are suitable for serving *à la vinaigrette.* Wash them under running water, immerse into cold, salted water, rub with a cut half of a lemon and dress with the *vinaigrette* given for leeks.

Artichokes. A fresh artichoke has firm, green leaves and is tightly furled. To prepare, cut the stem off close to the base and, should there be any withering outside leaves, remove them. Plunge into boiling, salted water slightly acidulated with lemon juice for 20-40 minutes, according to size. The very small ones can cook in as little as 10 minutes. To test if the globe is cooked, pull off an outside leaf and try the base of it with your teeth.

Serve cold with *vinaigrette*, or hot with melted butter.

Asparagus. The heads of fresh asparagus look like a flower in bud, tight and clean. Wash thoroughly under running water to get rid of the sand and, if there is time to spare, leave it to soak in plenty of clear water. Trim off the brown, jagged ends and lop any exceptionally long sticks down to the length of the average. Then, holding the asparagus gently just below the bud, scrape the rough coating off the stalk with a knife. Boil the asparagus, standing upright in bundles, in a tall, thin pan so that the stalks are in the water and the buds cook in the steam. (If you have no suitable pan, use the base of a double saucepan, reversing the top over the asparagus to hold in the steam.) The asparagus will cook in 12-15 minutes. Be careful not to overcook it. It loses much of its charm once it begins to droop.

Serve cold with *vinaigrette* given for cauliflower, or hot with melted butter or *Eliza Acton's Norfolk Sauce (see p. 186).*

Hors d'oeuvre is the one course of the meal which a busy woman can go out and buy from a delicatessen without feeling that she is detracting from her reputation as a cook (in fact, it is likely to enhance it because it leaves her free to concentrate her efforts on the main course).

The sultry tang of the smoke house is particularly pleasing to the palate at the beginning of the meal and smoked food makes the easiest of all *hors d'oeuvres* to prepare. Smoked sprats and smoked fillets of eel need only some slices of lemon; smoked trout, a little horseradish cream (cream whipped up with fresh grated horseradish and lemon juice); while smoked cod's liver *pâté* and smoked cod's roe demand only toast. Smoked ham or pork is good served simply with butter, or it can accompany melon or *crudities.*

An easy way of organising an *hors d'oeuvre* for an informal meal is to present a luxurious little row of fish in the Scandinavian fashion, some still in their tins, others laid out in small dishes, all arranged on a tray draped with a pretty napkin. The choice from the delicatessen is vast —

> **Fillets of mackerel in white wine**
> **Herring** in everything from aspic to a marinade of red wine or fruit juice
> **Sardines** or **pilchards**
> **Eel fillets** in oil or dill
> **Tuna in tomato**
> **Tinned Norwegian shrimps** or **prawns** —

to mention just a few.

Accompany the fish you choose with a few crisp, raw vegetables without dressing, spring onions, radishes or celery, or sweet, pickled cucumbers.

PRAWNS

If you can buy these freshly boiled (and not merely freshly unfrozen), they make a perfect and effortless *hors d'oeuvre* served still in their pretty pink shells. Allow a handful for each diner, and put a bowl of mayonnaise in the centre of the table.

* Leander Prawn Cocktail *

Prawn cocktail is one of the most predictable expense account starters, and a dish about which I have to admit to being *blasé*. But when dining in Henley recently at the Leander Club I was served with prawns dressed in a sauce about which it would be impossible to be *blasé*.

The steward, Mr. Cecil Smith, kindly gave me his recipe, remarking as he did so, "The sauce is not generally used, being more expensive than business will allow in most hotels." He is certainly right about its not being generally used.

For 1 pint of prawns allow ⅔ gill double cream, ⅓ gill Heinz tomato sauce, 1 tsp sherry and a generous dash of Worcester Sauce. Mix thoroughly and chill. The sauce is all the better for being made the day before it is needed. Shred a few lettuce leaves finely, place them in the bottom of a large wine glass, add shelled prawns (if you buy them freshly boiled and do the shelling yourself, the dish is twice as good) and pour on the cold sauce.

* Scandinavian Tomato Cocktail *

Clean tasting and packed with vitamins

Pour three-quarters of a tin of iced tomato juice into an electric blender. Add a spring onion and a stick of celery (both chopped) and season with salt, pepper and sugar. Blend until smooth.

Pour this mixture into wine glasses and put a bunch of watercress and the rest of the tomato juice into the machine. When the cress is liquidised, spoon a tablespoon of it into each glass. Serve with salt biscuits.

OYSTERS

I can think of no finer way of starting a meal than with a plate of oysters. Yet for some inexplicable reason the English, with their monopoly of the world's best oysters, so very rarely serve them in their own homes.

BUYING.   Today, although poverty and oysters are still poles apart, there are plenty within reach of anyone who does not regard a bottle of wine as an undue extravagance.

All the Mac Fisheries shops sell the firm's small Cornish oysters at 7s. 6d. for 12.   Some get a fresh supply of these family Helfords every day, others will order them for you given 24 hours' notice.

Better value, in my opinion, but less widely distributed, are the Portuguese oysters which some fishmongers sell.   Costing about 6s. a dozen, the creature inside the craggy looking shell is larger than its cousin from the Helford river hatcheries.

And the best buy of all, for a party, are the Portuguese "cocktail" oysters sold by the Whitstable Oyster Fishery Co. (39, Fish Street Hill, E.C.4) at 60s. a hundred, carriage paid.   (All prices are those prevailing in 1965.)

KEEPING.   Oysters can be kept *unopened* for 2 or 3 days in a cool larder or cellar — *but not in the refrigerator;* it would kill them and a dead oyster is a bad oyster.

Store them deep shell downwards covered with a damp towel or sack sprinkled with a few ice cubes and salt.   When you come to use the oysters, throw out any shells that are starting to gape or that open with suspicious ease.

OPENING.   Of course there is a certain knack to opening an oyster. It is easily learnt and is no more tricky than poaching an egg or pinning a nappy on a lively baby and I certainly hope that no one will be faint-hearted enough to get the fishmonger to open the oyster for them.

Because *oysters should never be opened more than 20 minutes (and preferably not more than 5) before a meal.*

To open an oyster, hold a folded cloth in the left hand and grasp the washed shell, deep side downwards and point towards you. Force a strong thin knife (preferably an oyster knife) horizontally between the shells at the hinge.   As soon as it penetrates swivel the blade on its axis so that as it turns towards the perpendicular, it forces the shells apart.

Cut the muscle which holds the fish to the flat shell, being careful not to spill the oyster's juice.   (This sounds much more complicated than it is, but any oystermonger will give you a demonstration.)

TO EAT RAW.   To eat raw (and it is sacrilege to eat natives in any other manner) serve the oysters in the deep shell bathed in their own liquor. A wedge (not a slice) of lemon, some brown bread and butter and a pot of cayenne pepper are the only proper accompaniments.

Guinness goes superbly with oysters — if they are making the meal. But if they are only the starter, a dry white wine is more suitable.

∗ *Fresh Salmon Paste* ∗

Given to me by Mr. True, the chef at the Imperial Hotel, Torquay, and slightly adapted

> 3 oz cooked salmon
> 1 gill double cream
> 3 tblsp whipped egg white
> Tabasco, ketchup, salt and pepper

Two salmon heads will give enough meat for this dish. Wash well before poaching them for 12-15 minutes in *court bouillon* made by boiling together for 25 minutes :

> 1 pint water
> 2 tblsp tarragon vinegar
> thyme, parsley, peppercorns, salt
> 1 onion, 1 carrot

When the head is cool enough to handle, pick off every scrap of pink flesh and pound it in a mortar before working it through a food mill twice (if you have an electric blender it will do all this for you, but you will have to put half the cream in with the fish to act as a lubricant).

Fold the paste into the lightly whipped cream (or what remains of the cream if you have used a blender). Fold in the egg white, season and turn the mixture into a small earthenware dish or *terrine*, and leave it to get thoroughly cold. Serve with freshly made white toast.

∗ *Kipper Pâté* ∗

> ½ lb filleted kipper (frozen fillets are excellent for this)
> ½ oz butter (for cooking the fillets)
> 1½ gills double cream (whipped)
> 1 tsp lemon juice

Tabasco, Worcester sauce, paprika and black pepper
but no salt
2 tblsp melted butter

Using a small covered pan, cook the kipper fillets very gently in
the ½ oz butter until soft enough to work through a food mill or
press through a sieve. If you use a food mill, put the fish through
twice. Fold the kipper *purée* into the whipped cream. Then,
tasting as you go, season generously before stirring in the remaining
2 tblsp butter — melted but not hot. Turn into a small *terrine*, or
tiny individual *ramekins* and leave in the fridge to set, for several
hours if possible.

* Potted Crab *

The principle of fish potting is simple, and fish of all sorts lend themselves to this delicious treatment — including trout, Dublin Bay
prawns, the delicate little cucumber-flavoured smelts, and, of course,
crab. This recipe can be adapted to any of these.

2 crabs (boiled)
4-5 oz butter
salt, pepper and powdered mace

Pound the white flesh of the crabs in a mortar till completely
smooth and season with salt, pepper and mace. Pile the flesh into
small oven-proof pots, cover with half the melted butter and bake
for ½ hour in a gentle oven, Regulo 3 or 335 deg. F. When cool,
seal with the rest of the butter.

* Ratatouille *

For 6-8 people

Although normally eaten hot, this rich *ragoût* of Mediterranean
vegetables makes a magnificent *hors d'oeuvre*, served cold and
dusted with fresh herbs. All recipes for *ratatouille* include four
basic ingredients — aubergines, tomatoes, sweet peppers and onions.
Choose deep purple aubergines with sleek skins, fleshy peppers and
strong onions.

A large heavy pan with a lid is needed for this dish because the
vegetables must simmer slowly.

2 large onions
2 peppers
2 aubergines
2 courgettes
4-6 tomatoes
12 black olives
2 cloves garlic
salt, freshly ground black pepper, ground coriander
9 tblsp olive oil
fresh basil (if possible) or parsley

An hour before you plan to start cooking, cut the aubergines and courgettes into slices, sprinkle with salt and leave in a colander to drain. Heat the oil and simmer the finely-sliced onions for 5 minutes before adding the peppers — cored, seeded and cut into slivers. After another 10 minutes add the aubergines and courgettes, and half an hour later tomatoes and cucumber (both chopped), creamed garlic and seasoning. Keep the lid on the pan all the time. Continue cooking for another hour, removing the lid for the last 10 minutes. Stir in the stoned olives and leave to cool. Drain off any excess oil.

Present the *ratatouille* at table in a large flat dish, generously sprinkled with chopped basil or parsley.

✳ *Fried Aubergines and Yoghourt* ✳

A dramatic little *hors d'oeuvre* (or a warming *entr'acte* in the middle of a cold meal) is crisply fried aubergines dressed with cold, garlicky yoghourt.

Dip the drained slices of aubergines in a light batter made with 4 oz flour, 1 egg, 1 gill water, and seasoned with salt, pepper and paprika. Deep fry in olive oil for 3-4 minutes (according to the thickness of the slices), and drain and serve at once with the yoghourt sauce — 1 gill seasoned yoghourt stirred into 1 creamed clove of garlic. If you do not want the bother or the fumes of deep frying, cook the drained aubergine slices in a frying pan with just enough oil for them to absorb and crispen.

✳ *Sweet Corn on the Cob* ✳

Worth eating only if it is tender, pale and plump. The sheath of leaves should be fresh, and the strands coming from the top soft and silky. A reliable test for young corn — dig your fingernail into an ear — if a milky liquid squirts out at you, the head will be good to eat.

If corn is young enough, 5 minutes in boiling salted water should be enough. Serve each diner with one head, plenty of butter (solid, not melted) and a spare napkin, as corn is eaten with the fingers.

* *Potted Pigeon* *

> 2 pigeons
> 6 oz butter
> 6 juniper berries (the berry that gives gin its name and flavour)
> nutmeg, mace and cinnamon — a generous pinch of each
> black pepper (freshly ground) and salt
> 2 rashers of bacon — each cut into three
> 1 tblsp of cooking brandy (optional)

Soften 4 oz of the butter and blend into it the juniper berries (crunched into granules), the spices and seasonings. Put 1 dessertsp of this spicy butter inside each bird and spread the rest liberally over their skins before placing them, breast downwards, in a small lidded casserole.

Add the brandy, cover the birds with a piece of buttered paper and place the casserole in a low oven, Regulo 2 or 310 deg. F., for at least 2 hours.

While the birds are still hot from the oven, strip the flesh off the bones. Pour the cooking liquor into a small bowl and leave it to cool until the butter solidifies on the top.

Line a small pot (I use a Stilton cheese jar) with 2 bits of the fried bacon, and pack in half the pigeon meat (finely chopped — not minced), pressing it firmly down. Repeat this process, topping the pot off with the last 2 bits of bacon.

Skim the butter off the cooking liquor, and scrape away any bits of meat or liquid sticking to its underside. Bring it just to melting point before pouring it over the potted pigeon, pressing the meat down again to ensure that the butter permeates right to the bottom. Melt the remaining 2 oz of butter and tip it over the surface of the meat to seal it from the air. Serve with butter and French bread or fresh toast.

Pigeon prepared in this way can be stored for 3-4 weeks in the refrigerator, provided the surface is resealed each time the pot is broached. The quantities given will provide an *hors d'oeuvre* for four, or lunch for two.

* Terrine of Grouse *

A way of turning an ancient grouse to good account is to break down its sinewy resistance with a mincer and turn it into a *terrine*. Another is to incorporate it, jointed, into a steak and mushroom pudding or pie.

> **2 old grouse (roasted for 25 minutes in the oven) plus their livers**
> **1 lb belly of pork and ¼ lb pie veal, minced by butcher**
> **¼ lb fat bacon cut into dice**
> **1 gill white wine, 2 tblsp cooking brandy**
> **garlic, crushed juniper berries, thyme, salt and pepper**

Strip the meat off the birds, mince it with the livers and mix in with the veal and pork, diced bacon, liquids (including any juices from the roasting pan) and seasonings. Line an earthenware casserole with streaky bacon before filling it with the meat mixture.

Cover the casserole, stand it in a pan of water and bake in a cool oven for 1½ hours. When it is nearly cool, weigh down with a heavy weight and leave the *terrine* to settle for several hours.

If you do not intend to eat the *pâté* at once, cover with melted lard and store in a cool place.

* Bourekakia *

Tiny pasties from the Middle East

For irresistible mouthfuls to accompany *apéritifs* it is hard to beat those tiny melting pasties at which the Turks and Greeks excel. And made just a little larger they are a delicious start to a cold summer meal.

The proper pastry for these savoury pasties (known as *filo* or *burek*) is so thin that it is almost transparent. It is something to buy, not make. Only Greek grocers sell *filo* paste, but the widely sold "Hammer Strudel" paste makes a perfect substitute.

If you cannot get it, or "Hammer Strudel" paste as a substitute, or prefer your own pastry, make a flaky paste (with 6 oz flour, 2 oz each butter and lard, a pinch of salt and 4 tblsp cold water), and then roll it out as thinly as possible.

Cut the paste into 2-inch squares (or, if using home made flaky, into oblongs as for tiny sausage rolls) and paint them with cool melted butter. Lay a teaspoon of one of the fillings given below near one corner of each square, bring the opposite corner over

(making a triangle) and secure the edges so that the filling cannot seep out.

Brush each one with an egg yolk beaten up with 1 tsp of water and ½ tsp of salt, and bake until golden on a greased baking sheet in a preheated oven at Regulo 4 or 355 deg. F., about 10 minutes for the bought paste, 15-20 for home made flaky.

FILLINGS

· 1 · ¾ lb cottage cheese
1 egg (beaten)
parsley and chives
salt, pepper and paprika

Work all the ingredients together until the mixture is perfectly smooth.

· 2 · 4 oz cooked, drained spinach
2 oz grated Parmesan
1 well-mashed anchovy
1 beaten egg
salt, pepper, nutmeg and origano

As before, make a smoothly blended mixture of all the ingredients, adding the egg last.

· 3 · 4 oz chicken livers
3 oz mushrooms
salt, pepper
garlic, parsley
1 oz butter for cooking

Fry the chopped livers lightly in foaming butter, add sliced mushrooms. As soon as they are cooked, add creamed garlic, seasoning and chopped parsley. Be sure the mixture is reasonably dry before putting it on the pastry.

SNAILS

I have never been able to understand why the English, who happily scoff mussels, winkles and whelks, wrinkle up their noses in disgust if anyone mentions edible snails.

S.T.C.B.—B

* *Escargots Bourguignonnes* *

This means snails prepared in the Burgundian manner, not snails native to the region. As there are simply not enough of the luscious black, vine-fed *escargots de Bourgogne* to go round even in Burgundian restaurants, most of the snails served are imported — some from Provence — the *petit gris* — but most from central Europe. Most of these are prepared *à la Bourguignonne* and either eaten fresh on the spot or packed into tins.

Tinned snails, which come into this country accompanied by their shells in a separate pack, are remarkably good vehicles for snail butter, which is the main reason for eating snails anyway (*see p. 195*). This must be freshly made as garlic quickly turns butter rancid.

All the English cook has to do is make the snail butter, put a little of it into each of the shells, push in the snail and seal the shell generously with more of the butter.

When all the shells are stuffed, set them *butter side up* on a baking sheet, or *gratin* dish sprinkled with rice or rock salt (unless of course you have snail platters), and bake for 5-10 minutes in a preheated oven at Regulo 4 or 355 deg. F. As soon as the butter starts to sizzle they are ready.

Most snail fanciers will have no trouble in putting away a dozen.

EGGS

"Of all the products put into requisition by the art of cookery", wrote Escoffier, "not one is so fruitful of variety, so universally liked and so complete in itself as the egg." "Eggs", states the editor of the latest volume to appear under the name of Mrs. Beeton, "are rich in protein, including ten amino acids considered indispensable for growth and tissue repair."

Could the contrast between the French and the English approaches to food be more neatly epitomised?

I suspect that one reason for our failure to exploit the versatility of eggs is the immutable association in the Englishman's mind between eggs and breakfast. Our four favourite ways of serving eggs, one could almost say our only ways, fried, poached, boiled and scrambled, are those least likely to clash with corn flakes and marmalade.

Another reason, perhaps, is the lack of suitable utensils, particularly individual *ramekins* or *cocottes* and the flat "eared" dishes for eggs baked *sur le plat*, made of metal, fire-proof china or oven glass. These little dishes are cheap, easily bought and have numerous uses in the kitchen.

The charm of eggs is their simplicity, but it is only the freshest of new laid eggs that qualify for the very simplest preparations — soft boiled in the shell, *en cocotte à la crème* or poached. Poaching, in fact, is physically impossible unless the egg is fresh because the white will not adhere to the yolk when it is dropped into boiling water. And those patent "poachers" do not poach the egg, they steam it.

* *Oeufs en Cocotte à la Crème* *

Properly done, which it seldom is, this is a truly delicate dish with which to grace the most elegant dinner. The preparation is simplicity itself.

Allow 1 new laid egg and 1 small *cocotte* for each diner, and have half an inch of water on the boil in a shallow roasting pan. Heat the *cocottes* by leaving them for several minutes in the water.

When they are heated through, put a knob of butter the size of a raisin in each, and as soon as the butter melts break in the egg. Cover the pan and leave it to simmer for 2-3 minutes, by which time the whites should be just starting to set.

Pour 1 tblsp of double cream over each egg, replace the cover and leave the *cocottes* in the boiling water for another minute or two. Serve immediately.

In summer time, when fresh herbs are available a few leaves of tarragon, basil or chives can be added with the cream.

If the water is boiling, and the empty *cocottes* are left to warm in the oven, a hostess can prepare this dish while her guests are making their way to the dinner table; but, as timing is all important, it is wise to stand over the eggs while they cook.

A simpler version of *oeufs en cocotte* can be made without cream. Merely leave the *cocotte* in the water undisturbed for 4-5 minutes until the white is set.

Cooked in this way the egg is ready to serve, seasoned simply with salt and pepper, or garnished with one or two asparagus tips (the tiny ones that are too small to serve on their own), a dessertsp of the clear juice left over from a roast, or a small sprinkling of minute *croûtons*.

* Stuffed Eggs *

Both these make easy and picturesque starters

·1·   GREEN . . . Hardboil as many eggs as there are diners (8 minutes in boiling water and 8 in cold). Shell the eggs, cut them in half and remove yolks. Mash these well before adding (for each yolk) 2 tsp each of cottage cheese and sieved spinach, 1 tsp double cream and a generous seasoning of salt, pepper and nutmeg.

Pile this mixture back in the egg, being careful not to cover the white as this would spoil its pretty green and white effect which is part of the attraction of the dish.

·2·   GARLIC . . . One way of introducing an undertone of garlic to a dish without offending the most parochial appetite, is to blanch the clove before it is used. The filling that follows would be utterly flat if the garlic were omitted, but it would take the palate of a *Michelin* inspector to pinpoint the source of its spiciness.

For each hardboiled egg yolk allow :
>   6 cloves of garlic
>   1 dessertsp softened butter
>   salt and pepper

Boil the eggs for 8 minutes and plunge them into cold water for 8. The shells will then come away easily and when the eggs are cut in half the yolks can be spooned out without breaking the whites.

Blanch the peeled cloves of garlic (drop them into boiling water for 60 seconds) before pounding them up with the egg yolk (a pestle and mortar make quick work of this job). Work in the butter and season lightly, then stuff the mixture into the whites from which the yolks were taken. To vary the filling slightly, a little parsley or paprika could be added.

* Oeufs Mollets *

Because it is so rare, unless one has chickens in the backyard, to get eggs fresh enough to poach satisfactorily, I nearly always substitute *oeufs mollets* (that is, eggs caught at the halfway stage between being soft and hardboiled) in dishes when poached eggs are normally required — including poached eggs on toast.

Plunge eggs first into boiling water for 5 minutes and then into cold for the same length of time. Tap them gently all over and carefully peel off the shell.

Serve *oeufs mollets* bathed in melted butter; or crowned with mayonnaise; or resting on a bed of buttered spinach (with or without a cheesy *béchamel*); or dusted with paprika and set on a layer of corn kernels; or on toast in place of a poached egg.

* Oeufs en Gelée *

Faced with the words aspic jelly most of us either turn the page or reach for the nearest tin of crystals. However, it is not very troublesome to make an excellent *aspic menager* which has all the flavour, if not quite the clarity, demanded by Escoffier. It may take a long time to cook but it demands no more attention than a sleeping baby. Anyhow, there is no point in making this delicate dish unless a natural meat jelly is used. Its charm lies in the blend of fresh eggs with the refined richness of the meat, cream and herbs. Gelatine simply will not do.

For enough jelly for 6 eggs, plus a bonus to store away for future braises and sauces, you need :

2 lb veal bones (chopped)
1 lb leg of beef (cut into squares)
1 calf's or pig's foot (cut in two lengthways)
2 onions (each stuck with a clove) and 2 carrots
1½ pints of water, 6 peppercorns, salt

Bring all the ingredients very slowly to the boil. Skim, cover and leave to simmer gently for 3 hours (or 40 minutes in a pressure cooker). Strain through a fine sieve and leave to cool before scraping off every last scrap of fat.

Measure out just over half a gill of this stock for each egg, add a pinch of dried tarragon and simmer for a few minutes. Add a dash of dry sherry (check the quantity by tasting and don't overdo it). Leave the stock to cool.

Meanwhile, prepare the *oeufs mollets* as described on the page opposite. Place an egg in each *ramekin*, cover with the cold but still liquid jelly poured in through a strainer. Leave to set. Before serving, float a thin layer of cream over the surface and sprinkle with chives or the finely-chopped tops of spring onions.

✳ *Egg Soufflé* ✳

Spectacular as either a starter or a main dish for lunch or supper
Prepare 4 *oeufs mollets*, lay them on the bottom of a buttered
7- or 8-inch *soufflé* dish (or large individual *ramekins* or *soufflé*
cases) and season with salt and pepper.

THE SOUFFLÉ MIXTURE
> **1½ oz butter**
> **1½ tblsp flour**
> **2¼ gills milk**
> **4 oz grated cheese**
> **(half Parmesan, half gruyère)**
> **3 tsp French mustard**
> **salt, pepper and nutmeg**
> **6 egg yolks and 7 whites**

Using a heavy pan (one large enough to hold the whole *soufflé*)
melt the butter. Remove from heat and stir in sifted flour. Let this
bubble for a few minutes without browning before gradually incor-
porating the heated milk, stirring all the time. When the mixture is
completely smooth, leave it to simmer on the lowest heat for about
10 minutes, stirring regularly. Then add finely grated cheese, beaten
egg yolks, and mustard, season generously. All this can be done in
advance.

Thirty minutes before dinner is to be served, heat the oven to
Regulo 7 or 425 deg. F., put the top shelf in the middle and place a
baking sheet on it. Whip the lightly-salted whites until they stand in
peaks, preferably with a bulbous wire whisk.

Blending the whites with the base is the most crucial moment in
the life of a *soufflé*. If the little bubbles trapped in the egg are burst
the *soufflé* will never rise to the occasion. Work the froth in, half at
a time, using a spatula and a gentle stirring movement. While it is
important that the blend should be even, do not try to smooth out
every last vestige of the whites. When all the whites are absorbed,
tip the mixture over the poached eggs, but do not pack the *soufflé*
dishes more than three-quarters full.

Set the dishes on the baking sheet and bake at once for 10-12
minutes.

* *Spinach Soufflé* *

A spinach *soufflé* is the quickest of the savoury *soufflés* to make because it does not require a *béchamel* base.   For ½ lb *puréed* spinach allow 3 egg yolks, 4 whites, 2 tblsp cream and seasoning. Stir the yolks into the *purée*, add the beaten cream, and season before folding in the whipped whites.   Cook in the centre of the oven preheated to Regulo 6 or 400 deg. F. for 12-15 minutes.

For variation a little chopped ham or anchovy may be folded into the spinach mixture.

-5-
LIGHT LUNCHES
AND SUPPERS

The dividing line between this chapter and the last is somewhat arbitrarily drawn. Many of the dishes included here would make magnificent starters to a serious meal — too magnificent, in fact, which is why I prefer to serve them as the star turn of a lesser meal, accompanied only by bread and butter, a salad and a glass of wine. However, I have asterisked those dishes which, served in small quantities, make particularly appealing first courses.

SCALLOPS

Before the war scallops were almost too cheap to be respectable. Now that they cost as much as good-sized native oysters, their social status has risen accordingly. However, while no one can contemplate less than 6 oysters, few people would want to consume more than 3 scallops.

For some reason it is customary in this country to sell scallops open. Perhaps the fishmonger feels that the poached egg appearance adds a decorative touch to his slab. There is no danger in buying an opened scallop, as there is with mussels and oysters, but it will lack the tangy sea-water flavour of the mollusc freshly opened just before it is to be cooked.

* *Scallops en Brochette* *

One of the simplest ways of serving scallops is threaded on a skewer and cooked under the grill.

For each scallop allow: 1 large white mushroom, 1 large rasher of bacon, seasoning.

Cut the white part of the scallop into chunks the size of a grape, keeping the coral in one piece. Cut the mushroom into 4 and the rasher into 4 or 6 slices just large enough to roll.

Thread up a skewer, starting and ending with a mushroom and alternating the rolled slices of bacon with bits of fish. Season with pepper, paint with melted butter and put them under the grill for 8-10 minutes (4-5 minutes each side). Serve on a bed of watercress.

Accompanied by a little buttered rice, 2 well-loaded skewers make a satisfying supper or lunch dish.

* *Baked Scallops* *

Cut the scallops into little chunks, lay them in the deep shell and season with salt, pepper and lemon juice. Sprinkle with breadcrumbs and dot liberally with butter.

Cover the shell with foil and put in the middle of a moderate oven, Regulo 3 or 335 deg. F., for about 25 minutes, for the last 5 minutes without the foil.

* *Scallops à la Meunière* *

Cut the white part of the scallops into slices the thickness of 2 half-crowns but leave the coral intact. Dip the pieces first into milk and then into seasoned flour before frying them very gently in a little hot, foaming, butter for 5 or 10 minutes.

Drain well and serve with a sauce made by adding a little more butter to what is in the pan already, and then raising the heat until the butter turns the colour of *Amontillado*. Add a dash of lemon juice and pour the bubbling sauce over the scallops. Garnish the dish with a sprinkling of parsley.

MUSSELS

Mussels put little strain on the purse and 2 quarts will make an ample supper meal for 2 or an *hors d'oeuvre* for 4 or 5. Yet in spite of their cheapness, mussels are well capable of understudying the lead part in many popular oyster hits (such as *angels-on-horseback, kebabs* and stew). But there is one part they can never play. *Mussels must not be eaten raw from the shell* (perhaps that is why they are called the *poor* man's oyster!). Another big difference is the opening ceremony. Where oysters are forced open a with knife, mussels are coaxed open with heat.

KEEPING. Put the mussels into a bucket of cold water as soon as they arrive from the fishmonger and leave them in a cool place until you are ready to prepare them. Do not put them in the fridge as this would kill them and from the cook's point of view a dead mussel is a bad mussel. For the same reason a gaping shell should be discarded when you start cleaning them.

CLEANING. Scrub each mussel thoroughly under running water, scrape it and beard it before putting it into a bowl of fresh water. Change this water 2 or 3 times, or leave the tap running slowly into it for 10 minutes or so.

OPENING. Chop up an onion, a shallot, a stick of celery, some parsley, thyme and half a bay leaf, and put this mixture into a heavy lidded pan with 1 tblsp of butter, and a gill of wine. Bring up to boiling point before throwing in the 2 quarts of cleaned mussels.

Cover the pan and leave it over a medium fast heat, shaking it at intervals. Within 3-5 minutes all the shells should be open. If any are still closed, give them another minute or two. Any that do not respond to this treatment should be thrown away. Now that the mussels are open they can be prepared in a variety of ways.

✳ *Moules Marinières* ✳

Having thrown out the empty half shells, transfer the full ones to a warmed casserole. Strain the liquor into a saucepan, add 1 gill of boiling cream. Season with black pepper, chopped parsley, and pour the liquid over the mussels. Serve at once.

* *Mussels Served with Snail Butter* *

Cover each mussel in the shell with about a teaspoonful of *beurre d'escargot* (*see p. 195*), arrange them in fireproof dishes, sprinkle with breadcrumbs and put them under a hot grill for a minute or two. (*see also moules en croûte, p. 152*)

OYSTERS

Some fanatics react to the notion of cooking oysters much as my husband might were I to take one of his *château* bottled clarets and tip it into a casserole.

And if the oysters involved are East Coast aristocrats then the fanatics are perfectly right. Oysters for cooking should be of *bourgeois* origin, and this is where Portuguese oysters and Mac Fisheries family Helfords come into their own.

* *Angels on Horseback* *

These are simply oysters moistened with lemon juice, wrapped in bacon, impaled on a skewer and grilled for 5 minutes, 2½ minutes each side.

* *Oyster Ramekins* *

For each *ramekin* allow :

> 6 Portuguese oysters
> 1 tblsp each white wine and cream
> 1 shallot, 4 mushrooms
> 2 dessertsp each celery and breadcrumbs
> 2 oz butter
> tarragon, parsley and cayenne pepper

Simmer the oysters for 30 seconds in a mixture of their own liquid and white wine. Leave them, off the heat, in the liquid until needed.

Soften the shallot, mushrooms and celery (all chopped) in 1½ oz butter with the tarragon. Put half the remaining butter in a *ramekin* and fill it as follows : breadcrumbs, vegetable mixture, drained oysters, vegetables, breadcrumbs.

Stir the cream into the oyster liquid, bring to boil and pour it over the breadcrumbs. Sprinkle the surface with parsley, a touch of cayenne, and top it with the last bit of butter. Cook for 6 minutes in a preheated oven at Regulo 6 or 405 deg. F.

* *Kedgeree* *

The English have always been good at breakfast, and some of our best national dishes are aristocrats of the breakfast table.

Now that hearty meals in the early morning have gone out of fashion, several of our more distinguished breakfast dishes, such as kedgeree and potted herring, have remustered to lunch or supper.

For kedgeree (an Anglo-Saxon *risotto* with Indian forbears) the transition was easy, especially as it is perfectly capable of holding its own with a good bottle of dry white wine.

$\frac{1}{2}$ lb patna rice
$\frac{1}{2}$ lb smoked haddock (frozen will do)
2 hardboiled eggs
2 oz butter
1 gill cream
salt, pepper, parsley

Simmer the haddock in salted water for 10-15 minutes. Remove all bones and flake up the flesh. Boil the rice in plenty of acidulated salted water, to which you have added the liquid in which the haddock was cooked, for about 10 minutes.

Drain the cooked rice and put it to dry for a few minutes in a cool oven before pouring over melted butter. Stir in flaked fish, most of the chopped hardboiled egg and finally the hot cream. Season with salt and freshly ground black pepper and sprinkle the surface with parsley and the rest of the egg.

* *Potted Herring* *

Put 1 gill white wine, $\frac{1}{2}$ gill each wine vinegar and water, into a pan with a few onion rings, 2 or 3 slices of carrot; also mace, peppercorns, fennel seed, bayleaf and parsley. Simmer for 15 minutes and allow to cool.

Have the herrings filleted, salt and pepper the fillets on both sides before rolling them, tail first and skin outwards, and laying the rolls in a casserole which will hold them snugly. Pour on the liquid before covering and baking for $1\frac{1}{2}$-2 hours at Regulo 2 or 310 deg. F.

* *Anchoïde* *

An *anchoïde* is *croûtes* (or toasts) spread with a *pâté* made from anchovies, oil and garlic.

It is a hole-filling snack which can easily be stretched to fill the gap between breakfast and dinner (or the other way round).

> 24 anchovies
> 2 cloves garlic
> squeeze of lemon juice
> pepper
> 1 tblsp olive oil
> 1 tblsp butter
> 8 slices of French bread toasted on one side only

Put the fillets of anchovy to soak in milk for an hour or so (or overnight) to rid them of excess salt. Pat dry, and then pound them with all the other ingredients until the mixture is a smooth paste. Spread this on to the half toasted bread, pop the slices under a hot grill for a few moments. Serve at once, accompanied by copious draughts of cheap red wine.

* *Leeks Sautéed with Bacon* *

For 1 lb of leeks, weighed after trimming, allow 4 rashers of bacon and 1½ oz butter. Using a heavy pan with a lid, lightly fry the diced bacon in the foaming butter. Add finely sliced leeks, salt and pepper. Stir thoroughly, cover and leave on a low heat for about 20 minutes. Stir occasionally.

This makes the most appetising supper dish on its own, or an ideal accompaniment for sausages or fried eggs. For a main meal try serving it with baked veal or grilled lamb chops.

* *Aubergine and Cheese Casserole* *

A fragrant savoury for lunch

> 1 aubergine (*prepared as described on p. 163*)
> 6 oz sliced gruyère (or mozzarella if you can get it)
> 1 lb tomatoes
> oil, salt and pepper, herbs and garlic

Lay some of the sliced aubergine in an oiled *soufflé* dish or casserole. Cover with a layer of cheese, followed by a layer of sliced tomatoes. Season with salt, pepper, basil or origano, or thyme and parsley, and chopped garlic. Repeat this process 2 or 3 times, ending up with a layer of cheese.

Bake, uncovered, for about 1-1¼ hours in a preheated oven at Regulo 3 or 335 deg. F.

* *Mozzarella in Carrozza* *

Literally cheese in carriages, this is a hot, lusty sandwich usually served with a fresh tomato sauce.

Make sandwiches with small slices of thinly cut white bread, from which the crusts have been removed, and a generous, well-fitting layer of *mozzarella* (a resilient white cheese — originally made of buffalo's milk — that comes bathed with its own butter-milk). If you cannot get *mozzarella*, *bel paese* makes an adequate substitute.

Sprinkle with seasoned flour and leave sandwiches to soak in a flat dish into which you have poured 2 well-beaten eggs. Leave for half an hour, turning them over once or twice.

Meanwhile, make the sauce (*see p. 187*).

Fry the sandwiches in hot olive oil for a few minutes each side until golden. Drain, cover with the tomato sauce and serve immediately.

* **Green Gnocchi* *

Gnocchi (Italian by origin) are tiny dumplings made of potato, semolina, cheese or spinach. The best of the four (in my opinion) are the spinach or green *gnocchi*. It is also the best way I know of eating spinach.

These delicate little mouthfuls have a flavour as evocative of Italy as a Puccini aria, but are little more troublesome to make than the English suet dumplings that traditionally accompany boiled beef and carrots.

My recipe for green *gnocchi* is based on that printed in *Italian Food* by Elizabeth David.

As an *hors d'oeuvre* (to be followed by something very plain such as cold cuts) the quantities given will serve four people, as a main course, two.

¾ lb raw spinach or an 11 oz packet of quick-frozen (Findus for
    choice)
4 oz cottage cheese
1 egg and 1 tsp butter
1 tblsp flour and 2 of grated Parmesan
salt, pepper and nutmeg

Clean and cook the spinach, press out every scrap of liquid and,
if fresh, chop finely. Return to pan, mix in mashed-up cottage
cheese and stir for about 10 minutes over low heat until the two
are perfectly blended.

Transfer mixture to a bowl, stir in butter, flour, Parmesan, seasoning and well-beaten egg. When the mixture is smooth, leave it to
stand for at least 2 hours.

While a large pan of salted water is coming to the boil, shape the
mixture into tiny sausages by dropping about a teaspoonful at a
time on to a floured board and rolling it over very lightly.

Drop the *gnocchi*, a few at a time, into the gently bubbling water.
After 4-5 minutes, when they rise to the surface, scoop them out with
a perforated spoon, drain well and lay them in a heated oven dish
containing melted butter and grated Parmesan. Keep warm while
the rest of the *gnocchi* are cooked. Dress with a little more butter
and Parmesan just before serving.

Green *gnocchi* accompanied by wholewheat bread and butter, a
green salad and a glass of *Bardolino* make a perfect lunch.

✳ *Potato Gnocchi* ✳

1 lb boiled and sieved potatoes
3 oz each flour and grated Parmesan cheese
1 beaten egg
salt and pepper

Knead the mixture until the ingredients are evenly blended.
Then, with floured hands, mould the paste into sausages the size of
a cork.

Meanwhile, bring a large pan of salted water up to boiling point.
When it is bubbling briskly, drop in the *gnocchi*, a few at a time.
As they rise to the surface, after 3-5 minutes, scoop them out with
a perforated spoon.

Give the *gnocchi* a moment to drain before laying them in a
heated casserole containing about 1 oz melted butter. Serve
sprinkled liberally with Parmesan cheese, or masked in a thick,

freshly made tomato sauce, or, if you can get hold of fresh basil, a *pesto* sauce (*see p. 192*).

* Savoury Pancakes *

For batter see p. 207–8

MUSHROOM FILLING

>   4 oz mushrooms, washed and sliced
>   1 hardboiled egg, chopped
>   1 clove garlic (optional)
>   parsley, thyme, salt and pepper
>   1 oz butter (scant)
>   2 tblsp sour cream (or fresh cream soured with lemon juice)

Sauté the mushrooms in butter flavoured with crushed garlic, thyme, salt and pepper. As soon as the mushrooms are cooked, stir in the sour cream, hardboiled egg, and sprinkling of parsley.

COTTAGE CHEESE FILLING

>   4 oz cottage cheese
>   2 or 3 pickled walnuts
>   stick of tender raw celery
>   2 spring onions or small bunch of chives
>   salt and freshly ground pepper
>   2 tblsp each of melted butter and grated Parmesan cheese

Finely chop the walnuts, celery and onions or chives and mix them into the cottage cheese. Season with salt and pepper.

Stuff the pancakes with one of these mixtures the moment they come out of the pan, brush with melted butter, sprinkle with cheese and set them for a few seconds under a hot grill. Serve immediately.

* Moussaka or Greek Shepherd's Pie *

This is a way of using up left-over lamb with *panache*. Strictly, it should be made with raw meat, of course, but this cooked meat version is worthy enough.

Chop the lamb you have left over as finely as possible, or put it through the mincer — though this is not as satisfactory. Moisten it in a frying pan with a little clear gravy or stock or water, and arrange it in alternating layers in a casserole with *ratatouille* (*see p. 61*) or the nearest approximation to *ratatouille* that is convenient. Start and finish with the vegetable mixture. Cover with a topping of yoghourt, egg and flour (1 gill yoghourt, 1 standard egg, 1 tsp flour seasoned with salt and pepper). Whip this together until the mixture is smooth, pour it over the *moussaka* and bake in a preheated oven at Regulo 6 or 405 deg. F. for 35-45 minutes. If the top has not browned, give it 60 seconds under a hot grill.

PASTA

A steaming plate of *pasta* makes a fine warming meal for a cold winter's day with the minimum of trouble. Meat and tomato sauces can, with advantage, be prepared the day before and heated up again while the *pasta* is cooking.

To boil spaghetti
Allowing approximately 3 oz for each person, boil unbroken, long spaghetti briskly in a large pan of salted water for about 15 minutes. Perfectly cooked *pasta* should be soft, but still offer a slight resistance to the teeth.

Turn the cooked, drained *pasta* into a hot dish containing a little olive oil to which a little garlic can be added, if you like, turning it over and over till each strand is shiny.

If you are generous with the garlic this treatment can replace the sauce entirely. It is certainly one of the most pleasant ways of eating spaghetti that is to accompany a *ragoût* of meat or fish. And with the addition of grated Parmesan, salad and a glass of wine it can make a very pleasant simple meal in itself.

* *Rich Meat Sauce for Spaghetti* *

4 oz stewing veal
2 oz pickled pork or smoked bacon or a mixture of the two
14 oz tin of Italian tomatoes
2 tblsp tomato paste
1 onion, 1 clove garlic
2 oz mushrooms (optional)
½ gill wine (optional)
3 tblsp olive oil and 1 of butter
salt and pepper, dried thyme, origano or marjoram

Using a heavy frying or *sauté* pan, soften the chopped onion and creamed garlic in mixture of oil and butter. Add the meat, finely chopped or minced, and raise the heat. When the meat has taken colour, drop in the sliced mushrooms and a minute later pour in the wine. Let it bubble fiercely and reduce by three-quarters before adding tomatoes, tomato *purée*, salt, fresh ground pepper and herbs.

Leave the mixture to simmer on a gentle heat (or in a slow oven) for 30 to 45 minutes, by which time all the watery liquid should have evaporated and the ingredients blended into a rich, thick and shiny sauce. A tablespoon of cream, fresh or sour, stirred in just before serving has a smoothing effect on both texture and flavour.

Other dressings for spaghetti are *pesto sauce (see p. 192)*, *mild tomato sauce (see p. 188)*, and *egg and bacon sauce (see p. 236)*.

✱ *Baked Lasagne* ✱

For 4 people

It is just as easy to make twice or four times as much if you want to serve this wonderful dish for a midwinter supper party.

>   5 oz lasagne or wide ribbon noodles
>   6-8 oz lean minced beef
> $\frac{3}{4}$ of a 14 oz tin of tomatoes
>   4 oz tomato paste
> $\frac{3}{4}$ lb cottage cheese
>   1 egg
>   2 oz grated Parmesan
> $\frac{1}{2}$ lb mozzarella, or if you can't get it, a mixture of gruyère and
>   cheddar
>   2 tsp dried origano and 1 of basil
>   parsley, salt, pepper, and 1 clove garlic

Brown the meat slowly in a little olive oil, then spoon off all excess fat. Add herbs, garlic, seasonings, tomatoes and tomato paste and leave it to simmer, uncovered, for about half an hour. Stir occasionally. Combine cottage cheese with beaten egg.

Meanwhile boil *lasagne* in a large pan of boiling salted water for approximately 20 minutes. Drain well.

Spread half the *pasta* over the bottom of a shallow 2-pint baking dish, about 3 inches deep. Cover with half the meat sauce, then half the cheese. Spread surface with very thinly sliced *mozzarella* (or *gruyère* and cheddar) and sprinkle with grated Parmesan. Repeat the layers.

If you do not want to eat the *lasagne* at once, at this stage it can be refrigerated. If you do that, add 5-10 minutes to the baking time.

Bake in a heated oven for about half an hour at Regulo 4 or 355 deg. F.

The *lasagne* will be easier to serve at a party if it is left to stand for about 10 minutes after it comes out of the oven.

* *Buttered Eggs and Smoked Salmon *

A luxurious dish that is quickly made and capable of starring on many occasions : at a grand breakfast, an elegant lunch, or an after-theatre supper. It also makes a fine starter to a dinner where the main course is either something cold or something simple like a grill.

The cheap, imported smoked salmon will do perfectly well for this dish — or the bits and pieces some shops sell off for a shilling or so.

> 5 eggs (as fresh as possible)
> ¼ lb smoked salmon
> ½ gill cream
> 1½ oz butter
> salt, black pepper, cayenne and parsley

Beat eggs with salt, pepper and 1 tblsp of the cream. Add smoked salmon cut in dice. Scramble the eggs in foaming butter, stirring constantly. Be careful not to overcook. Transfer immediately to a warm dish. Bring the rest of the cream to boiling point and pour it over the eggs. Garnish with chopped parsley and a touch of cayenne pepper. Serve at once.

* *Eggs Baked in the Spanish Style *

For each person

> 1 or 2 eggs (according to appetite)
> ½ medium onion
> 1 tomato
> 1 tsp butter
> 1 tblsp grated Parmesan cheese
> 2-3 tblsp double cream
> salt, pepper, and a touch of garlic

This is best made in individual *gratin* dishes, but one large dish can be used.

Melt the finely chopped onion in the butter, add sliced, peeled tomato and cook gently for a few minutes until the onion and tomato are blended. Add garlic, break in the eggs, season, sprinkle with cheese and cover with cream.

Cook for a few moments on top of the stove and then finish off under a hot grill for about 1 minute. The whites should be set but not solid.

Made with 2 eggs this becomes a delicious lunch or supper dish; made with 1, a substantial starter to a light dinner.

* *Eggs Baked in Peppers* *

For each person
> 1 large green pepper
> 2 eggs
> 2 dessertsp breadcrumbs
> 2 tsp butter, olive oil for greasing
> 1 tomato
> a touch of garlic (optional)
> freshly ground black pepper and plenty of salt

Cut pepper neatly in half, remove core, every last seed and any fibrous veins before dropping into boiling water for 6 or 7 minutes. Drain thoroughly. Paint the peppers inside and out with olive oil. Divide the breadcrumbs between the peppers, dot with half the butter, the chopped tomato, also a few drops of garlic juice and seasoning.

Break the eggs one at a time on to a saucer and gently slip them into the peppers. Dot with the rest of the butter and season again. Bake in a preheated oven at Regulo 4 or 355 deg. F., for 15 minutes.

If this is to be served as an *hors d'oeuvre*, halve the quantities and allow $\frac{1}{2}$ a pepper for each person.

* *Oeufs sur le Plat* *

Basically this means an egg (or several eggs) cooked in butter in a shallow dish, either on top of the stove or in the oven — but the possible elaborations are endless.

Allowing $\frac{1}{4}$ oz butter for each egg, put one big dish, or as many small ones as there are diners, into the oven preheated to Regulo 5 or 380 deg. F. As soon as the butter starts to froth, slide the eggs into the dish and season with salt and pepper.

Leave them to cook for 5-6 minutes — just long enough for the whites to solidify, but not long enough for the yolk to harden.

Eggs cooked in this manner will not wait upon the company, as the heat retained by the earthenware or metal dish will carry on the cooking process after the egg has left the oven.

The simplest garnish for *oeufs sur le plat* is a sprinkling of cheese added before the egg goes into the oven. Use *gruyère* or Parmesan, or a mixture of both.

Eggs dressed with snail butter are a refinement to delight garlic lovers (*for the butter see p. 195*). Slip a little of the mixture on to each egg just as it is to be served.

∗ *Tripes à la Lyonnaise* ∗

For 4 people
>   1 lb well drained cooked tripe
>   6 oz onions
>   2 oz butter

Slice the onions finely and cook them in the butter until they turn golden.

Raise the heat, add tripe cut into small strips and, turning it over and over in the pan, cook until it is all brown. Season with a little lemon juice and sprinkle with parsley. Serve at once, piping hot.

∗ *Trippa alla Parmigiana . . .* ∗

This is simply tripe cooked (*see p. 92*), dried, sliced, tossed in plenty of hot butter and dressed with lots of grated Parmesan cheese. Season with salt, pepper and nutmeg — and, for addicts, a touch of garlic.

∗ *Veal Kidney in Mustard Sauce* ∗

The recipe that follows is in the luxury class, but its preparation involves very little labour, and there is absolutely no waste.

>   1 veal kidney
>   4 oz mushrooms
>   2 spring onions
>   1 gill cream

salt, pepper, garlic, lemon juice
1 tsp fin de Dijon French mustard
1 tblsp butter

After stripping the fat off, cut the kidney into slices and let them colour in the melted butter. Add sliced mushrooms, spring onions cut into slivers, a few drops of lemon juice, salt and pepper and a touch of garlic. Cook on a moderate heat for 5 minutes or so.

Stir the mustard into the cream, bring the mixture to the boil before pouring it over the kidneys. Serve as soon as the sauce is blended with the juices in the pan, accompanied by some small slices of bread fried in butter, and buttered rice. Follow with a green salad.

–6–
CASSEROLES

All the dishes in this chapter can be made at least one day before they appear on the table. Some of them merely tolerate this treatment, others (notably those made with the tougher meats and containing wine, spices or tomato *purée*) actually taste better for being rested and reheated, which makes them particularly valuable to the busy woman who finds it suits her life to have a couple of grand weekly cookings rather than daily sessions at the stove. It also suits smaller families who are bound to see these large dishes make a second appearance (most casseroles are not worth making in minute quantities). After the first time of serving they can go away into the fridge for 3 or 4 days, only to reappear tasting better than they did originally.

• CLASSIC FRENCH CASSEROLES •

Many of the greatest dishes of French regional cooking are those warming farmhouse stews for which France is famous — *garbure, tripes à la mode de Caen, pot au feu, cassoulet,* all of them rich

89

dishes that would put heart into a hungry man on the coldest winter day.

One of the most satisfying of these, as well as one of the most easily composed, is the *cassoulet*. It can be simple or grand — according to the budget and the occasion — and it need lose little by being made in London instead of the Languedoc.

* *Cassoulet* *

Languedoc gave the word *cassoulet*, yet there is not a Languedocian living who could give you the approved, classic recipe — because it does not exist. Just as every coffee drinker has his own infallible recipe for the perfect infusion, so every Frenchman who cares for *cassoulet* knows just how it should be made. In practice the constituents of a *cassoulet* depend very much upon what the cook has in hand, except for the two invariables — haricot beans and some form of pork. Ham, bacon, lamb, poultry, game, sausages, herbs, wine — any or all of these, combined with beans and pork, can make up a *cassoulet*.

As the Languedoc is the land of the goose, one of the most regular "extras" is *confit d'oie* (preserved goose), while the fat most generally used for the preparation of the *cassoulet* (and every other imaginable dish) is goose dripping.

Making a *cassoulet* is simple enough. Basically it is an assembly job. Beans simmered slowly with a variety of flavourings (herbs, seasonings, vegetables and the rinds of pork or bacon) are combined with lightly baked meats and moistened with a rich, well-seasoned liquid. It is a dish that must be made in generous quantities; the amounts given here will provide an ample dinner for a family of 4 or 5 *twice over* (it is likely to be even better on the second time of serving), or be the mainstay of a midwinter supper party.

> 1 lb haricot beans
> 4-6 oz pickled pork
> 1 large onion stuck with 4 cloves
> 4 small onions cut in quarters
> bouquet garni, salt and pepper
> 2 cloves garlic
> 1½ lb pork spareribs
> 1 lb best end neck of lamb
> ¾ lb garlic sausage or boiling ring
> a leg and a wing of preserved goose (*see p. 91*), or any highly
>   flavoured roast bird

1 14 oz tin of tomatoes
2 tblsp fat (goose, duck or pork) for roasting
1 oz butter

Soak the beans overnight in cold water. Next day cover them
with fresh water and put them on to simmer for about 1½ hours
with the pickled pork, sliced up rind from both fresh and pickled
pork, boiling ring, *bouquet garni*, sliced garlic, salt and the onion
studded with cloves.

About half an hour before the beans finish cooking, set the pork,
lamb and small quartered onions to roast in the oven, preheated to
Regulo 5 or 380 deg. F. At the end of half an hour the meat should
be partially roasted and the beans soft and creamy.

Strain the beans, removing onion and *bouquet garni*, and cut all
the meat, both boiled and roasted, into convenient chunks. Place
cushion of beans at the bottom of a deep earthenware pot (of not
less than 4 pints capacity), cover it with chunks of meat, bits of
onion and joints of poultry.

Cover the viands with another layer of beans and continue in this
manner, seasoning lightly all the way, till all meat and beans have
been used, finishing with a layer of beans. Pour over the mashed up
tomatoes, and ½ pint of the bean liquor.

Sprinkle generously with breadcrumbs, dot the surface with
butter and put the *cassoulet* in a slow oven, Regulo 2 or 300 deg. F.,
for at least 1½ hours.

* *Confit d'Oie or Preserved Goose* *

1 fat goose (or duck)
a large pinch of ground cloves
2 crushed bay leaves
salt and lard

Get your butcher to cut the goose into about eight pieces. Strip
off all the fat and put it in a low oven to render. Mix salt with
cloves and bay leaves. Rub the surface of the joints of goose very
thoroughly with the seasoned salt mixture, pressing it in well. Leave
the meat to macerate in the salt for 2-3 days in a cool place.

At the end of this time wash and dry the joints, put them into
a large baking pan and cover completely with the rendered goose fat.
If there is not enough fat, make it up with pure lard. Cover the pan
and put it to cook in a slow oven, Regulo 2 or 310 deg. F., for about
1½ hours.

Test if the joints are cooked by pushing a skewer into the thickest part of the meat. If it slips in easily and the juice that squirts out is the palest pink, not red, the goose is cooked.

Drain the joints and place them in a glazed earthenware jar with a wide neck. When the cooking fat is cool, but still clear and liquid, pour it over the pieces of goose. Make sure that there is at least ¼ inch of fat between the meat and the open air.

Prepared in this way, and stored in a cool place, cooked goose will keep for as long as a year — provided that the air is always excluded by a layer of fat.

* *Tripe à la Mode de Caen* *

Tripe has not always been a music hall joke — a non-U comestible to be written off as one of the gastronomic foibles of the far-back Lancastrians.

In the robuster England of the Restoration, where appetites were less finicky, tripe was reckoned to be a dish for trenchermen of taste and fashion.

"Dined with my wife", Pepys told his Diary, "upon a most excellent dish of tripes of my own directing covered with mustard, as I have heretofore seen them done at my Lord Crew's."

It wasn't until the golden age of gentility, when Victoria Regina was on the throne and Isabella Beeton in the kitchen, that tripe came to be considered too vulgar for gentlemen to dine upon.

Of late, however, I sense a slight softening of the bearish market for tripe. Flanked by the more acceptable viands, it has been seen on the slabs of some of London's smartest butchers; while rumour says that the most expensive restaurants have been known to include the dish on the menu.

The only thing I have against tripe is the way we prepare it in this country. Instead of using it as a vehicle for other flavours, we either anaesthetise our palates by dousing it with vinegar, or over-emphasise its blandness by mashing it in a white sauce.

However, regardless of its ultimate culinary fate, when tripe arrives from the butcher the initial preparation is the same. It needs to spend 10 minutes under the cold water tap and at least 1½ hours simmering slowly in water flavoured with vegetables, herbs and spices, and invigorated with a dash of wine or wine vinegar.

Tripe à la Mode de Caen is the most splendid tripe dish of all. It could well be renamed from the point of view of its ingredients,

Tripe in the Somerset Manner. What marks its origins as indisputably French is the subtle blend of aromatic flavours harmonised by 8-9 hours' gentle simmering.

> 1½-2 lb honeycomb tripe
> 1 ox or calf's foot
> ¼ lb salt pork rashers
> 2 each carrots, onions and leeks
> 1 pint dry still cider
> 3 tblsp Calvados or Cognac
> bay, garlic, thyme, cloves, salt and pepper

Place 4 cloves on the bottom of a large earthenware casserole. Cover them with half the salt pork and half the chopped vegetables. Cut the calf's foot into chunks and the tripe (washed but not simmered) into squares and pile them on the vegetable bed.

Push the *bouquet garni* into the middle of the meat before covering it first with the rest of the vegetables, and then with the remaining rashers of pork. Pour over the cider and Calvados. Season well.

Cover casserole tightly and cook at bottom of a slow oven, Regulo 1 or 290 deg. F., for 8-9 hours.

✳ *Coq au Vin* ✳

Wine has two basic functions in the kitchen — as a marinade (for softening and flavouring meat, fish or poultry), and, as in *Coq au Vin*, as a cooking medium to enrich both the food cooked and the sauce surrounding it. The very cheapest wine will serve as an effective marinade, but something rather better is needed when the flavour of the dish depends on the flavour of the wine, in this case a sound Macon or Beaujolais costing around 9s.

For 4-5 people

> 3-4 lb roasting chicken
> 1 bottle Burgundy
> 4 oz salt belly of pork
> 2 oz butter
> 1 small onion
> 1 small carrot
> 2 tblsp brandy
> salt, pepper, bouquet garni, 1 clove garlic
> ¼ lb each button mushrooms and onions
> beurre manié (made by working 2½ tsp flour into 1½ tblsp butter)

Have the butcher joint the bird, and let the pieces soak all night in the wine. Melt half the butter and *sauté* the diced pork until it starts to crisp before adding the finely chopped onion and carrot. Let these colour slightly and then add the joints of chicken (thoroughly dried). When each joint has fried to a pleasant golden colour, pour over the heated brandy and set alight (to burn off the excess fat).

Cover with the wine in which the chicken was marinaded, add seasoning, *bouquet garni* and garlic. Cover the casserole tightly, and let it cook very slowly for 1½-2 hours.

Take out the chicken and remove the pan from the heat. Drop in raisin-sized pieces of the *beurre manié* — a few at a time. Stir vigorously. When all the pieces are melted, let the sauce bubble for a few minutes, put back the chicken and garnish with the button mushrooms and onions (a classic Burgundian garnish) *sautéed* in the remaining ounce of butter. Serve with *croûtons*.

✳ *Garbure* ✳

For 6 people

      8 oz salt pork
      8 oz boiling bacon
      6 oz garlic sausage
      2½ pints water
      1 each leek, carrot, pepper, onion, stick of celery
      ½ small cabbage
      2 cloves garlic
      parsley, marjoram, thyme

Using a large earthenware or enamel casserole, bring the 2½ pints of water to the boil. Meanwhile, blanch both pork and bacon. When the water starts to boil put in blanched meat, garlic sausage, and all the vegetables except the cabbage cut up small.

About 10 minutes after the water has returned to simmering point add the shredded cabbage, seasoning of salt, pepper, chopped garlic and herbs. Simmer slowly until the meat is tender enough to cut with a spoon — about 1 hour.

If you happen to have a *confit* in the larder (*see p. 91*), drop a joint or two with the fat still clinging to them into the pot shortly before serving. If you haven't, the remains of a roast bird or a few roast pork spareribs will work wonders. So will a few roast chestnuts, a common fruit of the region and one exploited to great effect by the cooks.

* *Daube* *

There are almost as many versions of a *provençale daube* as there are of a *bouillabaisse*. However, whereas each *bouillabaisse* recipe is treasured by its owner as THE authentic version, all others being heresies, cooks are prepared to admit that there is more than one authentic *daube*.

For 4-5 people
    1½ lb beef (chuck or topside)
    4 oz salt pork
    1 onion, 2 carrots
    1 gill stock or water
    2 cloves garlic, marjoram (dried will do)
    salt and pepper
    1 slice orange peel
    1 tblsp olive oil for frying

THE MARINADE
    2 tblsp olive oil, ½ pint red wine

Cut the beef into bite-sized chunks and roll it in a mixture made with the garlic and marjoram (both chopped), salt and pepper. Cover the meat with the oil and wine and leave it to marinade for at least 3 hours (or overnight).

Remove the meat, dry it and strain the marinade. Meanwhile cook the diced salt pork gently in 1 tblsp of oil, in a heavy lidded casserole, till it crispens. Add the onion and carrots (both chopped).

Boil the marinade briskly until reduced by a quarter, pour it over the meat and vegetables, add stock or water and a slice of orange peel. Cover the casserole, trapping the steam with a couple of sheets of greaseproof paper placed between the rim and the lid—or sealing it with a flour and water paste, and cook in a slow oven, Regulo 1 or 290 deg. F., for about 4 hours.

A little buttered *pasta* or some baked potatoes are the best partner for this robust *ragoût*, followed by a salad made with green peppers and mild Spanish onions cut in slivers and dressed with olive oil, salt and pepper.

* *Pot au Feu* *

This is, in fact, two dishes, a soup (the *bouillon*) and the cooked meat (*bouilli*), and it is only worth making on a scale that would serve 8 at a sitting or, of course, 4 people twice over. It is one of those comfortable, adaptable dishes which are almost no trouble to make and can feed the family for 2 or 3 days. You will need a gallon sized *marmite* or stockpot.

Choose one of the cheaper cuts of beef such as flank, silverside, brisket or blade bone steak—cut in one piece. Some people also like to include some shin of beef because it enriches the *bouillon*. If you do this, allow ⅓ shin to ⅔ of one of the other cuts. For 3 lb of meat allow :

> **2 sets of poultry giblets**
> **1 beef marrow bone**
> **6 pints water**
> **4 small carrots**
> **4 leeks (white part only)**
> **3 onions**
> **1 turnip**
> **1 parsnip**
> **a small head of celery**
> **a bouquet garni of parsley stalks, thyme, bay leaf**
> **salt (sea-salt for choice)**

If the butcher has not done it for you, tie the meat into a neat parcel with string and wrap the washed, chopped marrow bone in muslin. Put the meat and giblets, but not the bone, into the *marmite* with the water. Raise it very gently to boiling point and skim off the scum as it rises. When it ceases to rise, after about 20 minutes, add the cleaned vegetables (do not peel the onions), the *bouquet garni* and the salt. Cover loosely and cook for 2 hours so slowly that the water barely moves. If it boils, the *bouilli* will be stringy. Add the marrow bone and cook for another hour. To serve, remove the meat and keep it warm.

THE BOUILLON — Toast slices of French bread on both sides and spread it with the marrow from the bone. Serve in soup plates covered with *bouillon*.

THE BOUILLI — To eat hot, prepare some fresh vegetables to accompany the meat and serve with French mustard.

To eat cold, serve with pickled cucumbers and mild horseradish cream; or with *salsa verde* (*see p. 191*); or *mustard sauce* (*see p. 191*); or as a salad with *vinaigrette* (*see p. 181*).

It is not, of course, necessary to eat both the soup and the meat on the same day. The hot meat can be eaten the first day, and then the *bouillon* can precede the cold *bouilli* the second day.

* La Bourride *

Bourride, one of the famous fish *ragoûts* of Provence, is to my mind infinitely more enticing than the region's more widely publicised *bouillabaisse,* which I find a heavy and rather over-rated feast of fish.

Bourride is altogether subtler and less gargantuan. What's more, it can be made with the sterner fish that swim round the British Isles, whereas an authentic *bouillabaisse* cannot.

For 6 people

THE FISH

>About 3 lb of any 2 or 3 of the following : bream, mullet (grey or red), turbot, brill, John Dory, whiting, conger eel (cleaned and filleted)

THE STOCK

>1 quart water
>the heads and bones of the fish
>½ gill vinegar
>1 onion, 8 peppercorns
>1 slice orange zest
>bay, fennel, thyme

THE SAUCE

>¾ pint of aioli (*see p. 190*)
>2 egg yolks
>½ pint cooking liquor
>saffron
>12 slices French bread

To make the fish stock simmer the ingredients for ½ hour.

Dissolve a generous pinch of saffron in a little of the fish stock before covering the fish with the stock and the saffron. Leave it to poach for 15-20 minutes. When it is cooked, remove, cover with a little of the cooking liquor and keep warm.

Reduce the rest of the liquor by fast boiling to 1 pint.

To make the sauce, put the egg yolks into the top of a double saucepan and stir into it half the *aioli* (which can have been made the day before). Keep stirring all the time and take care not to let it cook too fast. Strain ½ pint of the reduced cooking liquor slowly into the egg mixture and stir constantly until the mixture starts to thicken.

Pour this immediately over the fried slices of bread which you have waiting in a large open dish and serve it at once as a soup. Follow with the fish accompanied by the rest of the *aioli*.

This dish is nothing like as complicated as it sounds and it makes a magnificent feast for anyone who enjoys garlic. After such a meal, no one will want anything more substantial than a salad, a little fruit and a cup of black coffee.

• A MISCELLANY OF MORE MODEST CASSEROLES •

These dishes are less ambitious than those in the previous group — the ingredients are cheaper and the recipes lend themselves to being made on a smaller scale.

* *A Provençal Casserole of Beef* *

Today's meat prices reflect the world shortage of meat, especially beef. There are, however, certain cuts of beef where prices are still at rock bottom because the butcher finds it hard to persuade the British public that, given a little trouble, fine meals can be made with cheap meat.

Two pounds of flank of beef, one of the cheapest cuts, will make a grand casserole for 4 people.

> 2 lb flank of beef
> ¼ lb salt belly of pork
> 1½ gills cooking wine
> ½ gill water
> olive oil
> thyme, parsley, marjoram
> ½ green pepper, 1 onion, 2 cloves garlic
> 1 tsp paprika
> salt, pepper and flour
> 2 tomatoes
> zest of orange
> a few black olives

Trim the excess fat off the meat and cut it into cubes. Put the beef and salt pork into an earthenware dish, add chopped onion, herbs, salt and pepper, and cover with the red wine. Leave to soak for several hours or overnight.

Remove the meat, pat it dry with kitchen paper, sprinkle with flour and brown it in olive oil. Add the wine in which the meat marinaded, water and all the rest of the ingredients except the olives.

Bring slowly to the boil, cover tightly and cook at the bottom of a slow oven, Regulo 1 or 290 deg. F., for 3-4 hours.

Leave to cool, preferably overnight, skim off any excess fat, bring back to simmering point and add the stoned black olives.

* *Oxtail Ragoût* *

For 4-6 people

Oxtail stew is the nearest thing in British cooking to a real French *ragoût* — a rich and subtle blend of cheap meat and vegetables left to mature slowly in the oven.

Soak the chopped tail for 4 hours, cover with fresh cold water and bring slowly to the boil. Skim, and simmer for ¼ hour.

Drain, dry and sprinkle generously with flour seasoned with salt, pepper, thyme and a touch of ground ginger. Melt 1 tblsp of pork fat in a heavy casserole and fry 3-4 oz diced belly of pork till the fat runs and the meat starts to crispen. Put in the oxtail and brown all over. Add a leek, an onion, a turnip, 2 carrots, 2 tomatoes (all chopped), plus 1 more small whole onion stuck with a clove.

Cover meat with stock, add a *bouquet garni* made with celery, parsley, bay leaf and a thin strip of orange peel, and season with salt, pepper and a crushed clove of garlic. Cover tightly and as soon as it starts to simmer, transfer to a slow oven, Regulo 1 or 290 deg. F, for 4 hours.

If you are not going to eat it until the following day, leave it to cool and skim off some of the fat. About 40 minutes before serving add a couple of handfuls of raw chopped vegetables to the pot, a leek, a couple of carrots, a small turnip, 2 or 3 sticks of celery, an onion. Simmer until they are just cooked but not mushy.

* *Carbonnades Flamandes* *

For 4 people

      1½ lb chuck steak
      3 oz gammon
      2 onions
      ¾ pint beer (preferably 2 parts light ale to 1 part brown ale)
      3 tblsp oil
      1 clove garlic, bouquet garni of bay, thyme and parsley
      salt, pepper, nutmeg
      1 tsp dark brown sugar
      1 tsp wine vinegar

Cut the meat into 2-inch squares, the gammon into large dice, and brown them both in the hot oil. Keep the browned meat hot while the thinly sliced onions are fried in the same fat, then arrange the meat and onions in layers in a casserole. Pour the beer into the frying pan and heat it slowly, scraping the edges of the pan with a spoon to release any bits which may have stuck on during the frying. Pour the hot beer over the meat and add the clove of garlic crushed with salt, the brown sugar, seasoning and *bouquet garni*. Cover tightly and cook for 3 hours in a slow oven, Regulo 1 or 290 deg. F. Stir in the vinegar and serve with plain boiled potatoes.

* Baked Pork Spareribs *

For 4 people
> 8 spareribs
> ¾ gill white wine
> juice of 2 oranges
> 1½ gills stock
> parsley, garlic, salt and pepper
> 3 crushed juniper berries (optional)

Trim some of the fat off the spareribs and, using a heavy *sauté* pan, set it in a moderate oven, Regulo 4 or 355 deg. F., to melt. Meanwhile rub the chops with a mixture of parsley, chopped garlic, salt, pepper and juniper. When the pork fat in the oven is running freely, transfer the pan to the top of the stove, remove the frizzled bits of fat and brown the ribs lightly on both sides.

Add the wine, let it bubble for a minute before pouring in the orange juice and the stock. Return the pan to the oven and leave it to cook uncovered, near the top, for about half an hour.

Serve with a platter of boiled rice generously dressed with melted butter, freshly ground black pepper, finely chopped raw onion and parsley.

This dish can be made in advance, or in double quantities, and reheated when it is needed.

* A Summer Casserole of Pork *

For 4 people
> 1 lb lean pork weighed without bone or fat
> 1 onion, 1 carrot, 1 stick of celery, 1 clove garlic
> 2 gills cider

salt, pepper, juniper berries, flour
1 gill sour cream mixed with 2 tsp flour
2 tblsp pork fat or olive oil

Colour the finely chopped vegetables in the fat, before adding pork cut in chunks, seasoned with salt, pepper and crushed juniper berries, and lightly sprinkled with flour.

When the meat is brown, pour on the cider, cover and simmer very slowly for 30-45 minutes (according to the cut used). Stir in the sour cream, simmer uncovered until the sauce starts to thicken and then serve with buttered rice.

* Osso Buco *

A shin of veal and a large tin of tomatoes — these are the basic essentials of *osso buco*, a rich and colourful dish from the plains of Lombardy.

It is simple and cheap enough for the most workaday dinner, and pretty enough for a party. The only difficulty is getting the butcher to give you the correct cut of meat. Insist on shin of veal, insist that the meat is not cut off the bone, and ask to have the shin cut into 2-inch slices. Ask for about 2 lb.

Fry these slices, mildly seasoned with salt and pepper, in 2 oz of butter until they start to colour, before adding an onion, a stick of celery, a carrot and leek. As soon as these start to brown, add 1 gill of white wine and let it bubble for a few minutes before pouring in a 14 oz tin of Italian tomatoes and approximately 1 gill of water or stock into which you have stirred 1 tsp of tomato paste. Add 1 tsp of dried herbs (basil or origano), cover and simmer for about 1½ hours.

Serve in a flat dish, covered in its own sauce (sieved if you are feeling fussy) and dressed with a mixture of chopped parsley and garlic (in the proportion of 1 handful to 1 clove) and the grated zest of ½ lemon.

The classic accompaniment to *osso buco* is *risotto milanese*.

* Risotto Milanese *

Making a *risotto* breaks all the conventional rules of rice cooking. Instead of being plunged into a copious pan of fast-boiling water, the rice is first *sautéed* in butter and then simmered in just enough liquid for the grains to absorb as they cook.

The rice becomes a vehicle for the flavours of all the other ingredients rather than acting as a foil or accompaniment to them.

For this reason it is important to use the oval translucent rice grown in the Po valley as it is so much more absorbent than the long-grained Patna. Italian rice is widely sold in this country now, but if you cannot get it, the stumpy looking variety from Java will do.

The possible variations on a simple *risotto* are endless, but the basic formula remains the same. The meat, fish or vegetables (or any combination of these) which will determine the character of the *risotto* are *sautéed* in the butter before the rice is added, and then cooked with the rice in the liquid.

The method and ingredients given below can be used for making a *risotto* with shellfish, mushrooms, or chicken.

For 3-4 people

> 1 cup Italian rice
> 2 cups well flavoured stock
> 1 gill white wine
> 1 onion
> 3 oz butter
> 1 oz grated Parmesan
> salt, pepper and saffron
> 1 tblsp of beef marrow (optional)

Put a generous pinch of saffron (or 3 or 4 filaments pounded to a powder) to soak in a few tablespoons of the hot stock. Meanwhile, using a heavy pan, let the finely chopped onion soften in 2 oz of butter. When it turns golden, add the marrow (if used), then the rice, turning it over and over in the butter until every grain is transparent and shiny.

Then add the wine and let it bubble until mostly evaporated before adding about ⅔ cup of boiling stock. Keep the rest of the stock simmering on the stove till needed. Let the rice simmer steadily, stirring occasionally, until most of the liquid has been absorbed before adding any more.

The total cooking time is about 25 minutes, though it may be just a little less. Add the saffron near the end of the cooking, and at this stage stir constantly as the last few minutes are crucial.

When the rice is cooked, season with salt and pepper, stir in the remaining butter and grated cheese.

The *risotto* should be creamy rather than granular. But do not interpret "creamy" as "mushy". The grains lose their independence with the Italian method of cooking rice, but they must not lose their identity as well.

* *Calf's Liver Casserole* *

For 4-6 people

1 oz salt pork
1½ lb calf's liver (in the piece)
2 carrots, 2 onions, 2 tomatoes and a stick of celery
bouquet garni of bayleaf, parsley, thyme, zest of lemon
2 tblsp sherry, ¾ gill each white wine and rich meat stock
salt, pepper, nutmeg
1 tblsp butter

Dice the pork and let it brown in a little of the butter. Add the rest of the butter, the cut-up vegetables, the liver (lightly dusted with flour). As soon as the meat starts to colour add the sherry and white wine. Let this bubble for a minute before adding the stock, herbs and seasoning. Cover the casserole very tightly and put it in a slow oven, Regulo 2 or 310 deg. F., for 1½-2 hours.

Just before serving, transfer the vegetables and meat to a serving dish, throw away the *bouquet garni*. If the gravy is too thin, reduce it a little by boiling before pouring it over the meat. Serve with small new potatoes buttered and lightly sprinkled with mint.

• CHICKEN IN THE POT •

Strictly speaking, not all these chicken recipes are casseroles in that the bird does not always come to the table in the dish in which it was cooked. On the other hand, they can all be cooked before they are needed, and require little or no last minute attention.

* *La Poule Henri Quatre* *

Take a large boiling bird, stuff it with a well-chopped mixture of its own giblets, a little bacon, a few ounces of fresh pork, perhaps a slice or so of smoked sausage if you have it, a thick slice of bread soaked in milk and squeezed dry, herbs, garlic, green pepper, celery, salt, pepper and a pinch of spice — all bound together with a beaten egg.

Using the largest lidded saucepan in your *batterie*, brown the bird all over in 2 oz of butter. Add whatever vegetables you have in the house (carrot, leek, turnip, onion, potato, celery, tomato — all or any of these, chopped), garlic, *bouquet garni*, seasoning and, if you are feeling lavish, some chopped knuckle of veal. Cover with boiling water, put on the lid and simmer for 2-3 hours.

If the bird is to be served as it is, about ¾ hour before the cooking is finished, remove the original vegetables and replace them with a good variety of fresh ones.

A refinement that makes a *chef d'oeuvre* out of this simple dish is a *vinaigrette aux oeufs* — in other words, an ordinary French dressing (5 parts oil to 1 wine vinegar plus the usual seasonings and herbs) to which the liquid yolks and chopped whites of 2 soft-boiled eggs have been added.

* Chicken with Rice *

Serve the chicken (as described in the last recipe) fresh from the broth — but jointed — accompanied by rice cooked as for a *risotto* in some of the liquor.

For 4 people allow 1 cup rice, 2 oz butter and 2 cups of broth. Toss the rice in the frothing butter, and as soon as it is translucent add the broth, a little at a time. After approximately 25 minutes the rice should be just cooked and all the liquid absorbed.

Surround the chicken joints with the rice to which you have added stoned olives, chopped tinned pimento, spring onions and celery — sliced.

* Chicken with Rice and Curry Sauce *

Serve the chicken with the rice (but without the final addition of vegetables) covered with a creamy curry sauce made with 1 oz butter, 1 tblsp flour, 1 tsp curry powder, 1 gill broth, ½ gill cream and seasoning.

Make this in the normal *béchamel* manner (*see p. 184*), adding the curry just after the flour and, stirring constantly, let it cook for 8-10 minutes on the lowest possible heat. Add broth, cream, seasoning and simmer for 10 more minutes.

* Chicken with Green Pepper and Yoghourt *

Prepare the rice as for *chicken with rice* above. Take 1 gill of the cooking liquor, reduce by a half before stirring in 1 gill yoghourt and 2 green peppers cored, de-seeded and chopped into slivers. Bring to the boil, season and pour over the hot chicken. Surround with rice and serve.

* *Paprika Chicken* *

Since most of us have to make do with broilers for most of the time, here is one good way of making a fine tasting and nourishing dish out of the comparatively flavourless flesh :

For 4-5 people
>3½ lb chicken
>½ gill each vermouth and white wine
>2 gills rich stock, 1 tin tomatoes
>1 onion, 1 green pepper, 1 clove garlic
>1 oz each butter, olive oil
>1 tsp paprika, 2 oz shelled walnuts
>salt, pepper, flour and bouquet garni

Joint chicken, roll in seasoned flour and brown in olive oil and butter. Add onion and allow to colour before adding vermouth and wine. Bubble for a few moments before adding hot stock, chopped pepper, creamed garlic, tomatoes, seasonings and *bouquet*. Cover and simmer for 20 minutes before adding chopped walnuts. Continue cooking for another ½ hour. Serve with rice or buttered noodles.

* *Poulet à l'ail from Béarn* *

In spite of the prodigious amount of garlic required for the following recipe, *Poulet Béarnaise*, the final effect is extraordinarily subtle and mild. By the time the dish reaches the table its pungency has evaporated. The chicken has acquired a bewitching flavour which is recognisably but not aggressively garlicky, while the cloves themselves—tender and almost sweet — are bathed in a blend of butter, natural gravy from the bird and their own juices.

For 4-5 people
>3-4 lb roasting chicken
>6-8 oz garlic
>3 oz butter and 1 tblsp olive oil
>salt, pepper and ¼ lemon
>½ gill giblet stock

Prepare the garlic by blanching the peeled cloves in boiling water for 30 seconds. Rub the cut side of the lemon over the chicken and sprinkle it with salt and pepper. Melt the butter with the oil in **a** large casserole, and let the chicken colour all over before putting in the blanched cloves of garlic and sitting it on top of them.

Spoon a little of the butter over the bird, and pour on the giblet stock.   Cover the casserole and transfer it to the oven heated to Regulo 6 or 405 deg. F., for 45-60 minutes according to the size of the chicken, removing the cover from the casserole half an hour before the bird is cooked.

* Gardener's Chicken *

For 4-5 people

    3-4 lb chicken
    ¼ lb mushrooms
    12 small onions
    2 oz rasher of streaky bacon
    ¾ lb small new potatoes
    4 tomatoes
    2 oz butter
    bouquet garni of thyme, parsley and marjoram
    ½ lb salsify or small young turnips or 6 artichoke hearts

Melt the onions and diced bacon in the butter, and keep them warm while the chicken is slowly browned all over in the pan. Return bacon and onions to the casserole, adding the sliced mushrooms, scraped potatoes, peeled tomatoes and the *bouquet garni*. Press a piece of tin foil well down over the chicken and replace the lid of the casserole firmly.   It is important that no steam should escape.

Cook for 1½ hours in an oven at Regulo 2 or 310 deg. F.   Have the salsify, turnips or artichoke hearts prepared and cooked, adding them to the rest of the ingredients half an hour before the cooking is finished.

Serve the chicken in its own gravy, surrounded by all the vegetables.

* Chicken en Cocotte *

For 5-6 people

    4 lb roasting chicken
    ¼ lb pickled pork
    2 oz butter
    12 small onions, 4 new carrots

¼ lb button mushrooms

1 lb potatoes, either tiny imported new ones or old ones cut to
   the size of a pecan nut, weighed after preparation

½ gill white wine, 2 tblsp brandy

bouquet garni, salt, pepper, lemon juice

Cut the pork into dice-sized squares and fry till coloured in melted butter, then add carrots, cut small, and onions, peeled but whole. When they start to colour, remove them and the pork and keep warm while slowly browning the bird in the same fat, having already rubbed it with lemon juice, salt and pepper. This part of the cooking will take about half an hour. When it is done, burn off the excess fat by pouring the warmed brandy over the bird and setting it alight. When the flames die down, add wine and let it bubble for a minute or two before adding herbs, potatoes and a little more salt and pepper. Cover tightly and cook in a very gentle oven for ¾-1 hour.

* *Chicken in Milk* *

Milk has two uses in the kitchen that are too rarely exploited — as a marinade and as a cooking medium for casseroling meat.

Bathing ingredients in a bowl of cold milk before cooking has the effect of drawing out unpleasantly powerful flavours. For instance, anchovies that are too salt or game that has become too gamey, both respond to the blandishments of milk by becoming milder to the taste.

On the other hand the meats that are most suitable for simmering in milk are white meats such as pork and veal, and of course chicken. One of the most effective ways of making athletic old game birds edible is to cook them slowly in milk for several hours.

For 4-6 people

   1 boiling chicken

   2 pints of seasoned milk

   4 oz mushrooms

   4 oz diced bacon

   2 onions

   butter for frying

   bouquet garni of celery, thyme and parsley

Prepare the milk by putting it in a pan with an onion stuck with 3 cloves, mace, bay, thyme, peppercorns and salt, bringing slowly up to boiling point and then leaving it to infuse for 30-40 minutes.

Dice the bacon and fry in the butter. Add chopped onions. Transfer to an ovenproof casserole. Cut the chicken into joints and fry them gently in the butter until golden. Put these on top of the bacon and onions in the casserole before frying the mushrooms in what remains of the butter and tipping them over the chicken. Add *bouquet garni*.

Heat up the milk again, pour it over the contents of the casserole and put this uncovered at the bottom of the oven, set at Regulo 2 or 310 deg. F. Allow about 2 hours after the milk starts to simmer.

Serve with buttered rice to which slivers of raw peppers and diced cucumber and celery have been added.

• GAME •

Grouse and partridge for the pot should be past their first youth — and not only for reasons of economy. A young bird that has not lived through its first winter does not have enough intrinsic flavour in its flesh to impart to the other ingredients in the casserole during the long, slow cooking.

* Old Grouse — Stewed in Milk *

For 2 people

> 1 large old grouse
> 1½ gills each milk and water
> 2 oz each mushrooms and belly of pork
> 2 rashers bacon and 1 tblsp fat
> bouquet garni, salt, pepper, mace

Cut the pork into small chunks and let it colour in the fat. Stuff the rashers into the bird, brown it in the same fat. Add the mushrooms and *bouquet garni* before covering the grouse with the heated milk and water. Season and cover tightly. Cook at Regulo 1 or 290 deg. F. for 3-4 hours.

Serve with small boiled potatoes and braised celery.

* *Partridge — with Sweet-Sour Red Cabbage* *

For 2-3 people

 2 stewing partridges
 1 lb red cabbage
 1 onion, 1 apple, 1 stick celery
 bouquet garni
 3 tblsp pork fat or olive oil
 3 tblsp cider or wine vinegar
 1 tblsp Demerara sugar
 salt, pepper, zest of orange

Roll the partridge in seasoned flour, and let it brown in some of the melted fat. Remove the bird and keep it warm. Melt the rest of the fat, add the vegetables and apple (all finely chopped) and turn them in the fat till they shine.

Put in the sliver of orange zest tied up with the *bouquet garni*, the vinegar, sugar and seasoning. Place the bird on top of the vegetables, cover the casserole tightly and put it at the bottom of a slow oven, Regulo 2 or 310 deg. F., for 2½-3 hours.

PIGEON

Pigeons, being officially classed as pests, are always in season.

Nearly all the pigeons offered for sale in this country are wood pigeons, most well past their adolescence and therefore unsuitable candidates for the flights of culinary fancy exercised by Escoffier and his like on the plump young squabs which are common in France. But, like most other creatures, what ageing pigeons lose in tenderness they gain in character. Few English pigeons are suitable for roasting, but they respond magnificently to the blandishments of gentle casserole cookery.

* *A Casserole of Pigeon* *

For 2 people

 2 pigeons
 2 oz belly of pork
 1 oz butter
 4 tblsp brandy
 1 clove garlic
 salt, black pepper and flour

THE MARINADE
> 1½ gills dryish white wine
> 1 chopped onion
> 1 chopped carrot
> bouquet garni

Season the cleaned birds with salt and pepper and leave them to soak in the marinade for several hours, or overnight, turning them two or three times.

When the birds are ready to cook, dice the pork and, using a heavy casserole with a lid, let it fry in three quarters of the butter until crisp. Dry the pigeons, sprinkle lightly with flour and gently brown them all over before dousing them with the warmed brandy and setting it alight (to burn off the excess fat). Add the strained marinade, creamed garlic and a little more seasoning. Cover tightly, putting a piece of foil between the casserole and the lid to prevent the steam escaping, and cook in the oven for 2½ hours at Regulo 1 or 290 deg. F.

Remove the birds and lay them on a bed of red cabbage and chestnuts (*see vegetable section, p. 168*).

* *Pyrennean Pigeon* *

For each person
> 1 pigeon
> 2 tblsp well flavoured chestnut purée
> 2 tblsp brandy (or 4 of wine)
> 1 oz butter
> 2 oz mushrooms
> salt and pepper
> 1 onion, 1 stick celery

Let chopped onion, celery and mushroom stalks colour in the melted butter before putting in seasoned pigeon. Let the bird brown before closing the pan tightly and cooking slowly for 30-45 minutes.

Add mushrooms and 5 minutes later pour the brandy or wine over the bird (I had this dish in the Armagnac country where this spirit is liberally employed) and continue simmering for another 10 or 15 minutes.

Meanwhile, prepare a *purée* of chestnuts. If you are pressed, unsweetened *purée* from a tin is all right if it is well flavoured and seasoned. Spoon the *purèe* into a heated casserole, sit the bird on top of it, surrounded by the mushrooms. Boil the juices in the pan briskly before pouring them over the pigeon.   Put the casserole into a medium oven for 10 minutes before serving.

HARE

Hares have provided inspiration for England's poets as well as her cooks ever since the days of Piers Plowman.   Certainly jugged hare is one of our few contributions to the world of *haute cuisine* — though Hodgson's "Little hunted hare" and Blake's searing image of the hunted creature's agonised outcry may have spoiled many people's appetite for this fine, rich dish.

However, such scruples need not inhibit modern gourmets. These days most hares are shot.

A young leveret can provide a magnificent meal for 8 people, or 2 completely different dishes for 4 — the saddle roasted and the rest casseroled or potted.   It is the cheapest luxury I know.

English hares make better eating than their smaller blue Scottish cousins.   The young animal is recognisable by its sleek fur, soft pads and thin rounded ears which are easily torn.

Never buy a hare that you have not seen first in its fur.   The butcher will dress it and divide it into saddle for roasting and joints for the pot.

However it is to be cooked, the texture and flavour of a hare are improved by a bath in an aromatic marinade.   It has the effect of both tenderising the beast and enhancing its flavour.   And it is not wasted because the marinade is later used as the cooking liquor for casseroled hare and provides the sauce for the roast saddle (*see p. 129*).

* *Marinade for a Hare* *

1½ cups cheap red wine (I use Algerian)
½ gill olive oil
1 sliced onion
bay, thyme, marjoram and parsley
6 crushed juniper berries (if possible)
2 thinly pared strips of orange zest
black pepper

Put all the ingredients except the oil into a pan and simmer for 5 minutes to release the aromatic oils in the herbs, spices and oranges.   As soon as they are cool add the oil and pour ⅔ of the marinade over those parts of carcass earmarked for the casserole, and the rest over the saddle. Leave in a cool place for several hours or overnight.

* *Casserole of Hare* *

    the legs of hare in the marinade
    2 oz belly of pork
    1 onion, 1 carrot
    4 oz mushrooms
    flour, salt and pepper
    possibly a little more liquid — water, wine or stock

Cut the pork into squares and set it in a heavy pan to melt. Remove legs from marinade, dry and roll in seasoned flour. Then, with the pork golden and the fat running freely, add the hare. When each joint is well browned on both sides, transfer the meat to a deep casserole. Dice carrot and onion and slice mushrooms, browning in the fat still remaining in the pan before adding to the meat in the casserole.

Pour the strained marinade into the pan and scrape it round with a wooden spoon to loosen all the bits of meat adhering to the base. When the liquid is boiling, pour it over the meat in the casserole. Transfer to a slow oven, Regulo 1 or 290 deg. F., for 2½-3 hours. Serve with a *purée* of potatoes slightly flavoured with orange.

FISH

Eels are the perfect raw material for a *matelote*, either by themselves or with a selection of coarse fish. I prefer the eels by themselves.

* *Matelote of Eels* *

For 4-6 people
    1½ lb eels, cut into 2-inch hunks
    2 onions, 1 clove garlic
    2 tblsp butter, 1 tblsp oil
    ½ bottle white wine
    carrot, bouquet garni, mace, salt and pepper
    12 each button onions and mushrooms
    1 egg yolk and 2 tblsp cream

Colour one of the onions, chopped in butter, before adding the eel. When the pieces are golden, pour off any excess fat before covering with wine. As soon as wine reaches simmering point, add carrot, the remaining onion, creamed garlic, *bouquet garni* and seasoning and cook, covered, for 25 minutes.

Remove fish and keep warm while thickening the sauce by adding a little of it to the mixture of egg yolk and cream before tipping them both back into the cooking pot. Heat until the sauce thickens, *but do not boil.* Pour over the eel, garnish with the *sautéed* button mushrooms, the blanched and *sautéed* little onions, and some triangles of bread fried in butter.

* Stuffed Baked Herrings *

2 herrings
yolks of 2 hardboiled eggs
1 soft roe
a little grated lemon zest
1 oz butter
salt, pepper and parsley
$1\frac{1}{4}$ gills white wine or cider

Pound the roe and egg yolks together until they form a fine paste before working in the softened butter and all the seasonings. Prepare the herrings, removing backbone and filling cavities with the stuffing. Secure the opening with cocktail sticks. Lay the stuffed fish in a buttered oven-proof dish, pour over the white wine or cider and bake in the oven at Regulo 5 or 380 deg. F. for 20 minutes.

* Paprika Plaice or Cod with Sour Cream *

4 fillets of plaice (ask the fishmonger to remove the black skin
   when he fillets the fish), or 4 small cod steaks
1 small onion, $\frac{1}{2}$ green pepper
2 tomatoes
$\frac{1}{2}$ oz butter
1 gill sour cream or cream soured with lemon juice, or yoghourt
salt, pepper and $\frac{1}{4}$-$\frac{1}{2}$ tsp paprika

Spread the butter on the bottom of a small lidded casserole and cover it first with a mixture of finely chopped onion and green pepper and then a layer of sliced tomato. Season lightly.

S.T.C.B.—H

Lay the fish on top of the vegetables, adding a touch more salt and pepper. Stir the paprika into the sour cream and pour it over the fish.

Cover the casserole and place it in the centre of a moderate oven, Regulo 4 or 355 deg. F., for 20 minutes.

* *Sole Bourguignonne* *

For 4 people
>   2 soles, whole or filleted
>   2 gills red wine
>   2 oz small mushrooms
>   24 small onions and 2 shallots
>   bouquet garni of thyme, parsley and a bay leaf
>   beurre manié (2 tblsp butter worked into 1 tblsp flour)
>   2 oz butter
>   salt and pepper

While lightly cooking the little onions and mushrooms in butter, cut the shallots into paper-thin slices and put them on the bottom of a shallow ovenproof casserole. Season the sole with salt and pepper, lay it on the shallots and surround it with the onions and mushrooms. Pour the wine into the pan in which the vegetables were cooked, bring almost but not quite to the boil, and pour it over the sole. Put in the *bouquet garni* and cook the whole in an oven preheated to Regulo 4 or 355 deg. F. for 20-25 minutes. For fillets 10 minutes would be long enough.

When the fish is cooked, drain off the sauce and thicken it with the *beurre manié*. Pour the sauce over the sole and return it to the oven for a few minutes or put it under the grill. Serve with *croûtons*, small cubes of bread fried in butter until crisp.

* *St. Peter's Pie* *

>   ½ lb cod fillet
>   ½ lb smoked haddock fillet
>   2 hardboiled eggs
>   1 clove garlic
>   2 tblsp butter melted in an equal quantity of olive oil
>   ¾ pint milk and 2 tblsp cream
>   1 oz butter
>   1 tblsp flour

2 oz grated cheese: for choice equal quantities of Parmesan and gruyère
salt, pepper, nutmeg, mace and bay

Spice the milk with salt, pepper, mace and bay leaf and bring it to boiling point before pouring it over the fillets.

Cover the casserole and place it in a moderate oven, Regulo 4 or 355 deg. F., for half an hour.

When the fish is cooked, strain off the cooking liquid, and pour ½ pint into a saucepan for the cheese sauce. Keep the rest in reserve.

To make the sauce, melt the butter in a small heavy pan and as soon as it froths move the pan from the heat and stir in the flour. Add just enough of the warm milk to the *roux* to form a thick paste.

Replace the pan on the heat and slowly stir in the rest of the milk. Season with salt, pepper and nutmeg and leave the sauce to simmer very slightly for about 15 minutes while making the rest of the preparations, not forgetting to stir it every few minutes. Don't add the cheese at this stage.

Meanwhile, flake the fillets, carefully removing every scrap of skin and bone, into a large bowl and mash them up with a fork.

Gradually work into this mixture the melted oil and butter, the cream, the crushed garlic and 3-4 tblsp of the milk.

When all the ingredients are perfectly blended adjust the seasoning and transfer the mixture to a shallow pie dish and cover it with slices of hardboiled egg.

Fold the grated cheese into the sauce, pour it over the fish and put the pie into a hot oven, Regulo 7 or 425 deg. F., for a few minutes — just long enough for the top to start to colour.

-7-
THE
SHOCK OF HEAT

• FISH •

Grilling is the simplest and most primitive method of cooking fish
It is also the way in which the true flavour of the fish is most effectively revealed. For this reason only the freshest of fish are suitable
candidates for the shock of the grill, there being no protective
batter or other enveloping sauce to camouflage its shortcomings.
The same strictures apply, of course, to the roasting of fish. However, if you know a good fishmonger, it is still possible, occasionally,
to get almost fresh fish. These are usually either herring, mackerel,
sole or other fish caught in off-shore waters and shipped immediately
to market, or coarse fish such as cod brought in from the distant
fishing grounds on one of Britain's small fleet of factory ships
equipped with freezing plant. Instead of making the usual 2-6
weeks' journey packed in ice, cod landed from the factory ships is
frozen within minutes of catching.

The best accompaniment to fish grilled or roasted is butter
flavoured with mustard, anchovy, herb or paprika. (*Recipes for
these will be found on pp. 194-5.*)

GRILLING

Remember, the thicker the fish the gentler the heat and the longer the cooking time.

* Herring and Mackerel *

Both can be enjoyed at their best grilled. Have the fish cleaned and beheaded and, if you can persuade him, ask the fishmonger to bone the herring. Wash and dry the fish thoroughly. Season with salt and pepper, stuff it with a little butter and make one or two incisions on either side. Heat both grill and grid, paint the grid and the fish with oil and cook the fish for about 4 minutes on each side. Serve the herring with a mustard butter or just plain French mustard; the mackerel with an anchovy or herb butter (fennel would be an excellent herb for this).

* Red Mullet *

Have the fishmonger remove the gills of the mullet but nothing else. Mullet taste better if cooked complete with their innards. Make three incisions in each side of the fish and leave it to soak in a bath of oil made aromatic with thyme, tarragon, fennel or parsley, and seasoned with salt, pepper and lemon juice. Grill as for herring or mackerel, allowing at least 8 minutes for each side.

* Cod *

Don't try this unless you are sure the cod is fresh. Allow 1 steak for each person. Season with salt, pepper and paprika and lay them in a dish containing olive oil for at least half an hour, turning them over once in that time. Remove the grid from the grill pan, melt a little butter in the pan and cook the steaks under the heated grill for about 10 minutes, turning once, but not again after the first 60 seconds, using a moderate heat and adding a few tablespoons of water as the juices start to dry up. Serve with a paprika, anchovy or garlicky green butter.

* Sardines *

Very occasionally fresh sardines can be found on the slab of an enterprising fishmonger. Richards (11, Brewer Street) of Soho have them more often than anyone else. Although sardines grilled in this country will never taste quite as delectable as they did at that little café in the Algarve or the Côte d'Azur, they will taste good enough to make the search for them worth while. And if you have a charcoal-burning barbecue stove in the shed, this will be the time to bring it out.

Grill the fish whole, lightly seasoned, allowing just a minute or two on each side. As soon as the skin starts to wrinkle and crispen they are ready to eat. Serve immediately. Allow at leasι 6 for each person.

* Lobster *

Lobsters must be cooked as soon as they are killed
The main problem with lobsters, other than the price, is killing the creatures. Very rightly, it is against the law for a fishmonger to sell an uncooked lobster unless it is alive, which leaves the cook to do the dastardly deed. There are a multitude of views on which way a lobster dislikes least meeting its end. Both the instant death in boiling water and the gradual onset of unconsciousness caused by raising the heat of the water slowly have their partisans, amongst cooks, that is, not lobsters. I believe the kindest way of despatching a lobster is to plunge a sharp instrument into the soft part of its body between the head and the tail. Out of consideration for the lobster I won't attempt to describe this process, but a fishmonger will teach you.

Cut the newly killed lobster in half, paint it lavishly with a mixture of melted butter and good olive oil, and place it under a heated grill as far from the heat as possible. Turn it over every 5 minutes or so and baste frequently. Serve with more melted butter. It will take 25-35 minutes in all. To my mind, this is quite the best way of eating lobster.

ROASTING

As far as fish are concerned, this is a loose term to cover baking in a moderate oven, or high temperature roasting on a grid.

* Oven Baked Cod *

Use either a whole codling or a cut of a larger fish. Season with pepper before wrapping the fish in thin rashers of streaky bacon. Lay it in a buttered casserole and surround it with small onions and slices of potato (partly cooked by frying in butter). Bake in the centre of a moderate oven, Regulo 5 or 380 deg. F., allowing 10-12 minutes' cooking time per pound. Drain off excess bacon fat before serving, and sprinkle with chopped parsley.

HIGH TEMPERATURE ROASTING

This is suitable only for larger fish, weighing more than 2 lb.

Make incisions in the side of cleaned (but not beheaded) fish and leave them to marinade in oil as described for cod steaks. Lay them on a well-oiled grid (or oven shelf) in the centre of a preheated oven and bake for 12-15 minutes per pound at Regulo 7 or 425 deg. F. When the skin is crisp and golden the fish is ready.

• MEAT •

SPIT-ROASTING

The reason why. Like bread and wine, roast meat is one of the oldest and most basic foods of man and one of which he never tires. But what we today loosely call a roast is, in fact, a bake from the baking oven, because, so far, not even the most sophisticated of space age ovens can simulate the dry, airy roasting heat of an open fire or an exposed electric element. In the chemistry of the roast beef of old England one essential factor is invariably lacking, air, or, to be more exact, oxygen.

The demise of spit-roasting was regretted by anyone who ever had the luck to discover just what he was missing. Which may explain why, after an interval of nearly 100 years, spit-roasting with a gas or electric substitute for the "nice clear fire" decreed by Mrs. Beeton is enjoying a renaissance in the land that gave the world roast beef.

It is tempting to dismiss the current vogue for spit-roasting as just another ephemeral craze, sparked off by the manufacturers of spit-roasters — a pretty gimmick latched on to by food writers in search of something new to write about. In fact, the pressure has been the other way round; the demand, small at first, but growing all the time, came from the eating public.

Now several modern cookers (from four gas and eight electric cooker manufacturers) come complete with a spit roasting attachment for fixing under the eye-level grill, while Cannon Industries have gone one better. In addition to the spit roasters built into their more expensive cookers (both gas and electric), they market a portable electric combined spit roaster and grill, costing in 1965, £31 10s. 0d., and available through electricity showrooms. This is large enough to roast a turkey and slim enough to slip into a suitcase for country weekend use.

One of the justifiable claims made for spit-roasting (apart from the flavour of the food) is that it roasts without leaving a legacy of laborious scouring. The cause of spluttering in the oven, as well as that pervasive "all-round-the-house" smell that accompanies the baking of meat, is the temperature of the fat in the baking pan. As the cooking proceeds, cool juices ooze out of the meat and drop into the hot fat, making a series of minute explosions which deposit a residue of mess on the walls of the oven and send an emanation of fatty smells around the kitchen.

How. Spit-roasting does not involve any startlingly new techniques. The basic rules are much the same as for oven-roasting, and all the recipes that follow (except for *gigot provençale*) can, with advantage, be executed on the spit, including the hare and the saddle of lamb.

As with oven roasting, beef should be sealed quickly under a strong heat which is then reduced, while white meat such as pork and chicken responds to more even heat.

All spit-roasted meat is better for basting and usually it needs just a little longer than it would have done had it been cooked in the oven; a chicken weighing 3 lb after cleaning requires about 50 minutes in the oven, or 60 on the spit.

Lean meat needs larding, dry fleshed birds require wrapping in layers of pork fat (barding).

At the time of writing modern guides to spit-roasting are almost non-existent. Apart from the brief primer on the subject given by Cannon to their customers, the only book devoted to this subject is more inspirational than practical. By Sylvain Clusells, a great chef and a master spit-roaster, *Cooking on Turning Spit and Grill* is a craftsman's book.

The following list of do's and don'ts is drawn from my own experience :

Do not expect the impossible from a spit. It will do wonders with fine materials, but it cannot make a frozen battery hen taste like a fresh, free-range spring chicken.

Never be tempted to shorten the cooking time by raising the heat. And if in doubt about the intensity of heat required, always err on the side of moderation.

It is important to spear the joint or bird plumb through the middle so its weight is evenly distributed. Also be sure the meat is properly centred on the spear.

Do not salt meat or poultry until cooking is nearly complete. It hinders the browning process.

Do not try to spit-roast veal, the flesh is too dry.

When roasting, keep a little hot stock or water at hand to add to the meat juices in the drip-pan as soon as they start to caramelise.

To make the gravy after the meat or bird has been removed, pour off the excess fat, if any, and add a little more liquid, wine, stock or water, to the juices in the drip-pan. Raise the heat and stir vigorously.

Most meat needs boning for the spit. However, leg of lamb can be roasted on the bone, as can saddle. The shaft of the spit fits down the spine of the rather large lambs on sale in this country.

When roasting pork, try threading apples on the spit for the last 20 minutes of the cooking time. Baste them with pork dripping.

Just before serving pork, stop the spit from turning when the meat is fat side up and switch the heat up to maximum just long enough for the crackling to break into blisters.

A corner of gammon can be cooked alongside a chicken — giving the gammon a head start over the bird.

When the family is small and you do not want to put on the oven just for roast potatoes, they can go on the spit. Par-boil 3 or 4 large ones for 10 minutes, paint with oil or dripping, then thread them on either side of the roast.

Never be afraid to experiment; 19 times out of 20 you will be pleasantly surprised.

LAMB

In view of the universal popularity of roast meat, I have never been able to understand the peculiar prejudice which appears to exist against serving a roast, especially lamb, at a dinner party — unless, of course, it is something upstage like a game bird; nor the reluctance to present roast meat with any dressing other than the usual gravy made of flour and water, and, of course, mint sauce and horseradish, when there are so many simple but superb alternatives.

Early English lamb, at its innocent best, needs no adornment other than its own juices diluted with a little white wine or water, and mint butter. The slightly gamier lamb from the Commonwealth can stand more robust treatment.

Whatever roasting device you employ, lamb prepared in the Greek manner makes a pleasant change. Leg or saddle can be spit-roasted on the bone, shoulder must be boned.

Rub the meat well with lemon juice before making a few slits in the flesh and working into them a mixture of crushed garlic, chopped thyme (dried will do) and black pepper. Paint with olive oil and roast in the usual way, basting frequently with more oil slightly sharpened with lemon juice.

If the meat is spit-roasted, allow 30-35 minutes for each pound; in the oven, 25 minutes a pound is enough. Salt the joint about 15 minutes before serving.

* Coriander Lamb *

This recipe and the following one can be applied to either a leg or shoulder of lamb.

Paint the meat with olive oil and rub it all over with a blend of ground coriander and crushed garlic, making a few incisions on the surface and pressing the aromatic mixture well into the grain of the meat. Roast in the ordinary way, sprinkling the joint with salt about 20 minutes before serving.

* Spring Lamb *

Roast the meat in the usual manner (in butter for choice). Ten minutes before it is ready to be served, sprinkle it with a mixture made with equal quantities of fresh breadcrumbs and chopped parsley seasoned with fresh ground black pepper and (optionally) a little chopped garlic.

Pour a little more hot butter over the meat, and return it to the oven, raised to Regulo 8 or 445 deg. F., just long enough for the dressing to brown, which should take about 10 minutes.

* Gigot à la Provençale *

Roast leg of lamb in the style of Provence, is a perfect marriage of English method and Mediterranean flavours. The manner of roasting is exactly the same as for a conventional English roast, except that it is slightly slower and takes rather longer. For a 3 lb leg allow at least 2 hours.

3-4 lb leg of lamb
2 cloves garlic
2 onions
1 large aubergine
4 tomatoes
6 black olives
½ green pepper
olive oil
salt, pepper and herbs

Cut 1 clove of garlic into slivers and push into small incisions made in the surface of the meat. Paint the joint with olive oil and set it in a preheated oven set at Regulo 7 or 425 deg. F. After 10 minutes reduce the heat to Regulo 4 or 355 deg. F. Baste.

Chop the onions and put them under the meat to cook. In about 20 minutes baste the meat again and then spread the bottom of the roasting pan with the aubergine (previously cut into slices, sprinkled with salt and left to drain). Leave the aubergine to cook for approximately 45 minutes, turning the slices once in this time.

Add finely sliced green pepper, and season both meat and the vegetables beneath it with salt and pepper. Let another ½ hour go by before adding skinned, chopped tomatoes and the second clove of garlic very finely cut up. A few minutes before serving stir in the stoned olives (cut into pieces) and chopped herbs.

Serve the joint on a large platter surrounded by its rich, wine-coloured sauce. It will need no gravy or green vegetable. Just a *purée* of potatoes to accompany it and a green salad to follow.

* *Saddle of Lamb* *

It consists of both loins of the lamb from the ribs downwards and weighs between 4 and 7 lb, so this is a joint for large gatherings only. For smaller households, *carré*, which is actually half a saddle, is more practical.

The flavour of saddle is such that it needs no improving. Just roast it in a moderate oven for 20 minutes to the pound (25 for spit-roasting), using butter to lubricate the meat. Fifteen minutes before the joint is cooked, season with salt and pepper. Serve with its own juices diluted with wine or water and with *rowanberry jelly* (*see p. 226*) or a *béarnaise sauce* (*see p. 185*) lightly flavoured with fresh mint or thyme. Mint sauce, in my opinion, is an insult to the meat, the wine and the cook.

BEEF

Of all cuts of any animal the most luxurious is *contre-filet*, the eye of the sirloin, cut in the French fashion and rolled.

* *Roast Contre-Filet* *

This is an expensive cut wherever it is bought, though you can eat every ounce you pay for and a little feeds a lot of people (2½ lb being enough for six). However, it is slightly less costly when bought from butchers such as Benoit Bulke of Old Compton Street in Soho, or Harrods, who are accustomed to cutting meat in the continental fashion. When you buy the meat, make sure that a thin layer of fat has been wrapped round it before it is tied up with string for roasting. A red wine sauce (*see p. 186*) would be a nice complement to this elegant roast.

Remove the meat from the fridge at least two hours before cooking. Place it on a rack, give it a generous coating of melted butter and place it, unseasoned, in the centre of a hot oven, Regulo 8 or 445 deg. F. After 10 minutes reduce the heat to Regulo 5 or 380 deg. F. Baste the meat every 10 minutes and allow 12 minutes' cooking time for each pound of meat. So a 3 lb roast would take only 45 minutes in all (50 on the spit) — for those who like their beef rare. About 10 minutes before the meat is cooked, season generously with salt and pepper.

When the meat is ready, transfer it to a really hot carving dish. Pour off all the fat, loosen the meat juices in the pan with 1 tblsp of hot water and tip them into the red wine sauce (if you have made one — this can, of course, be done the day before).

Serve with green butter (*see p. 193*) and the sauce.

CHICKEN

There are numerous ways of making the best of tender but tasteless factory raised birds, but roasting is not one of them. This revealing treatment should be reserved for properly fed, free-range chickens which, in response to growing public demand, an increasing number of butchers and poulterers are stocking. Unfortunately, until it arrives on the table, there is no way of distinguishing between a broiler and a naturally reared bird, so you must be able to trust your butcher.

When you do get (and pay the premium price for) a first-class chicken, do not spoil the farmer's good work by roasting it in any old fat, or mixture of fats, which you happen to have handy. A good bird deserves butter.

Season the bird inside and out with lemon juice, salt and pepper. Push some butter and a few fresh herbs or mushrooms into the cavity. Heat the oven to Regulo 6 or 405 deg. F., melt more butter in the roasting pan before putting in the chicken on its side. Turn the bird over every 15 minutes, basting regularly until after the third time of turning it is sitting upright. A 3-4 lb chicken will be cooked after 45 minutes in the oven, 50-55 on the spit. The combination of butter and chicken juices should provide all the gravy that is needed; if not, stretch it with a little giblet stock, white wine or water.

* *Capon* *

These can be cooked in the same way, allowing 20 minutes for the first pound and 15 minutes for the remaining pounds. Serve with cucumber sauce (*see p. 186*).

* *Tarragon Chicken* *

Tarragon has a remarkable affinity with chicken and *courgettes*.

> **2-4 lb roasting chicken**
> **¾ lb courgettes**
> **4 oz butter**
> **several sprigs of tarragon**
> **salt, pepper and zest of lemon**

Push a large sprig of tarragon, ½ oz of butter, zest of lemon and seasoning into the chicken and roast in the usual way in the rest of the butter. Halfway through the cooking put ¾ lb *courgettes*, sliced longways, into the roasting pan.

When the bird is cooked serve it surrounded with the mixture of butter, chicken juices and *courgettes* seasoned with chopped tarragon and a little more salt and pepper. Accompanied only by a few new potatoes and a green salad, this version of tarragon chicken provides the perfect heart to a delicate meal.

If using a spit for roasting chicken, cook the *courgettes* in a separate pan using the butter that has accumulated in the drip pan under the chicken.

* *Roast Duck with Wine, Orange and Honey Sauce* *

Choose a fresh duck with a plump breast, soft down, pliable beak — and weighing at least 3 lb. The high proportion of bone to flesh makes ducks weighing less than this particularly extravagant.

Stuff a 3 lb duck with 2 small peeled oranges and a stick of celery, seasoned with salt and pepper. Prick the bird all over, rub with salt, and set it on a roasting rack in a hot oven, Regulo 8 or 445 deg. F., for 10-15 minutes, then reduce the heat to Regulo 5 or 380 deg. F.

After half an hour at the reduced heat, spoon off most of the fat in the roasting pan and pour $\frac{1}{2}$ gill each of white wine and orange juice over the duck. Baste the bird regularly with this mixture.

Fifteen minutes later spoon 2 tblsp melted honey over the duck; after another $\frac{1}{4}$ hour repeat this process with just 1 tblsp of honey. Fifteen minutes later the duck should be a shiny golden brown and ready to serve. To make the gravy, skim off any excess fat, add a little more wine and orange juice, boil briskly for a minute or two. Allow an extra 10-15 minutes if the bird is being cooked on a spit.

GAME

Game once figured prominently in the Englishman's diet. In the days before cheap chilled meat was imported from abroad and poultry was mass-produced, it was a valuable source of fresh food, being at its best just when meat was most scarce. It also brought a welcome variety into the monotonous meals of the winter months.

Even today, much game is not expensive. Hare, a fine meaty creature, is magnificent value for money; venison is rich and cheap, while pigeons, which make a good meal for two, seldom cost more than half a crown.

At a more luxurious level, woodcock are certainly worth snapping up if you have the chance. So is blackgame, a relation of the grouse, provided due allowance is made for the natural dryness of the flesh.

* *Woodcock* *

Woodcock are roasted undrawn sitting on a slice of bread fried in butter. Skewer the long beak through the body of the bird, cover it with a rasher of fat pork or bacon and roast for 15 minutes in a preheated oven at Regulo 6 or 405 deg. F.

* *Roast Blackgame* *

Stuff the bird with about 6 oz seasoned steak, spread the breast and legs with butter and drape the bird with thin slices of fat pork or bacon. Put it to roast in an oven preheated to Regulo 5 or 380 deg. F for ¾-1 hour, basting frequently.

Halfway through the cooking, spoon 2 or 3 tblsp of red wine over the bird. Serve with creamed potatoes tinged with a touch of grated orange peel, and chicory baked in butter.

* *Roast Wild Duck* *

Game birds that come all too rarely to the table are members of the wild duck family, mallard, widgeon and teal.

It is important not to overcook wild duck — its flavour is as elusive as its name suggests. A perfectionist once wrote that "wild duck should fly through a fierce oven and be served immediately". As I have yet to meet the bird prepared to co-operate in such an operation, I always compromise with perfection and use the following recipe.

Stuff the bird with half a segmented orange and 1 oz butter, sprinkle with salt and pepper, and paint with melted butter. Lay the bird on the grid at the top of a hot oven, Regulo 8 or 445 deg. F., with a drip pan on the shelf immediately below.

Cook for 15-20 minutes, basting frequently. Five minutes before the bird is ready, pour over it 2 tblsp of port diluted with 2 tblsp orange juice.

Serve on a bed of watercress, accompanied by redcurrant or rowanberry jelly, and the gravy in the drip pan diluted with a little giblet stock.

Sometimes in hard winters wild duck feed on marine weed close by the sea shore. Although some have claimed that this makes them barely edible there is a simple remedy recommended by Elizabeth David in *French Country Cooking*.

"Put 1 tblsp of salt and an onion into the cavity of the birds . . . and put boiling water to the depth of a quarter of an inch into the baking pan. Bake for 10 minutes, basting the ducks with the water." Drain and roast in the ordinary way.

* Guinea Fowl *

In recent times, these mildly gamey birds have been more popular in France than they have over here, but they can now be found at an increasing number of poulterers. This is due to the enterprise of a Scottish soldier turned farmer, Major Tom Adam, founder and proprietor of Denmore, Britain's only guinea fowl farm. Guinea fowl taste more like pheasant than chicken and look like miniature speckled turkeys. Whenever Tudor menus, or Shakespeare, mentioned turkey they were referring to guinea fowl. If your butcher cannot get them for you and you are not within shopping distance of stores such as Harrods, Selfridges or Whiteleys, write to A. S. Juniper, 369 Central Market, London, E.C.1.

They can be cooked in any method suitable for pheasant, but this is my favourite recipe :

> **1 guinea fowl**
> **1 lb white grapes**
> **butter**
> **salt and pepper**

Stuff the bird with half the grapes. Smear it generously with butter and put into a preheated oven at Regulo 7 or 425 deg. F. After 10 minutes reduce the oven to Regulo 5 or 380 deg. F., basting constantly. Allow 45 minutes cooking time in all. Ten minutes before the bird is ready, baste it with juice squeezed out of half the remaining grapes. Skin and seed the remaining quarter and add them to the juices in the pan just before dishing up the bird. Serve this grape sauce separately, or carve the bird into joints before sending it to the table and pour the sauce over it.

HARE

For a description of the preparation of a hare *see p. 111.*

* Roast Saddle of Hare *

> **1 saddle (with or without the back legs)**
> **½ gill cream**
> **2 or 3 thick rashers of fat bacon for larding**
> **2 tblsp pork fat or butter for cooking**
> **salt and pepper**

S.T.C.B.— I

Preheat the oven to Regulo 7 or 425 deg. F. so that it heats while the hare is being prepared.   Remove the saddle from the marinade and dry well.   Cut the bacon into inch long strips (about $\frac{1}{8}$-inch wide) and work them into incisions made with a sharp pointed knife, or, using a larding needle, thread the lardoons, as these strips are called, into the saddle, allowing about 6-8 stitches each side of the spine.

Season the saddle with freshly ground black pepper, place it on a grid in a baking pan, or straight on the shelf of the oven with a pan to catch the drips underneath, and pour over it the hot pork fat or butter.

Reduce the heat to Regulo 5 or 380 deg. F. and continue cooking for an hour, basting regularly.   Bring the strained marinade up to boiling point, pour it over the saddle, and continue cooking for another half hour, basting at least twice.

Pour the cream (previously brought to boiling point) over the hare, and baste once again before transferring it to a dish.   Let the sauce in the drip pan simmer for a few moments on the top of the stove, stirring thoroughly.   Serve separately.   On the spit allow another 10 minutes.

* *Fondue Bourguignonne* *

For cook-it-yourself occasions

Each member of the family, or visitor, deep fries his own steak to his liking, a mouthful at a time, in a pan of hot oil set in the centre of the table on a spirit stove, and dresses it with one of a variety of sauces made by the cook beforehand.   If you are serving it to guests, warn them in advance.   Fondue parties are lots of fun but not an occasion for dressing up in vulnerable clothes.

There are any number of fondue sets to be found in the shops, some very attractive, nearly all rather expensive.   In fact, no special equipment is needed.   A deep saucepan or cast-iron casserole set on a trivet and warmed by a spirit lamp works quite well.   However, each diner must have a long handled fork with a wooden or insulated handle (or a long wooden skewer) and a separate fork for eating the meat after it has been fried — if he is not to burn his mouth.

Allow about 6 oz beef (or pork) fillet for each diner.   Trim the meat of all fat and connective tissue, cut into bite-sized cubes and season lightly with black pepper.   Present each diner with his portion of meat uncooked.   Fill the fondue pan not more than $\frac{1}{3}$ full of olive oil and let it heat gently until a crumb of bread, when dropped into it, sets up a sizzling and starts to colour.   When the guests are gathered round the table each spears the meat and cooks it in the hot oil — a cube at a time. With the meat serve hot French bread and butter; sweet pickled cucumbers; raw celery; a dish of onion; hard boiled egg and parsley and a selection of the following sauces, choosing those which contrast most effectively with each other :—

> mustard sauce
> horseradish
> barbecue
> hot tomato
> mild tomato
> *béarnaise* (or one of the mayonnaise based sauces — tartare, aioli, *sauce verte*)

Recipes for all these can be found in the sauce chapter.

If you settle for three, choose one that is bland (like a *béarnaise*), one that's spicy (the barbecue, for instance) and one that is clean and cool (such as the *tartare*). And if you wanted to complete a quartet, add the sour cream and mustard which combines the virtues of all three.

–8–
COLD FOOD

Cold food does not necessarily mean summer food but (except for cold cuts) it almost invariably means rather grand food. This is partly because cold dishes, being more static than hot ones, need to be more carefully composed, and partly because the palate is automatically more critical when there is no heat to distract it or emphasise the flavour of the food. (Our increased sensitivity to the flavour of cold food is a point to be remembered when adjusting seasoning.)

This is not to say that cold food is inferior to hot; on the contrary. Nearly all fat meats, pork, duck and goose, for instance, are every bit as good cold as hot, and to some tastes, superior, provided the meat is of good quality in the first place. While turkey (to my mind) is hardly worth eating hot, the meat of a well fed and freshly killed bird can be excellent cold.

Most meats destined for the cold table will be both moister and better flavoured if oven baked in the French manner rather than the high temperature roasting which is customary in this country.

The instructions that follow are for French roasting of chicken, but the same rules can be applied to pork, duck, turkey and veal, but not to beef which needs the short, sharp shock of heat to seal the exterior while leaving the centre red.

∗ *French Roasted Chicken* ∗

- **·1·** Make a stock with the giblets.
- **·2·** Stuff the bird with 1 or 2 oz butter and either tarragon or rosemary.
- **·3·** Season with salt and pepper, inside and out.
- **·4·** Spread the outside with butter and cover with a piece of greaseproof or foil.
- **·5·** Place the bird on a rack in the roasting pan and fill the pan with the stock till it reaches the parson's nose.
- **·6·** Bake in the centre of a preheated oven, Regulo 5 or 380 deg. F., allowing 20 minutes to the pound.
- **·7·** Baste with the liquid several times and turn the bird round once.
- **·8·** Fifteen minutes before the chicken is cooked remove the paper so the breast can brown.

Chicken cooked in this way can, of course, be served hot, in which case the liquid in the pan will serve as the gravy.

∗ *Lamb* ∗

Lamb is not a meat which I would deliberately set out to serve cold, simply because it is so very much nicer hot. However, it is no penance to eat cold lamb if the meat has been cleverly seasoned before cooking with chopped herbs (rosemary or thyme and marjoram or parsley), creamed garlic, salt and pepper. Stuff this mixture into about a dozen little incisions made in the surface of the meat, allowing about 1 dessertsp for a 2½ lb leg.

··· ∗ ···

Four rather grand dishes for private celebrations or summer dinner parties. All are somewhat extravagant with time and ingredients but all have the advantage of making no last-minute demands on the cook.

∗ *Ballottine* ∗

For 6-7 people

A *ballottine* is a boned, stuffed bird encased in its own jelly. It is a dish as impressive as it sounds, yet it is as simple as a sausage to carve and, in spite of all the words it takes to describe, it is easy enough to prepare once you have persuaded your butcher to co-operate by doing the boning.

3-3½ lb chicken
1 calf's foot and a few veal bones
2 onions, 2 carrots and 2 oz mushrooms
2 tblsp cognac
1 oz butter, salt and pepper

THE STUFFING
6 oz each pork and veal
2 chickens' livers and 2 oz finely chopped fat bacon
1 shallot and 1 stick celery, finely chopped
2 tblsp each melted butter and cognac
garlic, parsley and a pinch each of thyme and bay
1 beaten egg
salt, black pepper

The day the butcher plans to bone the bird, ask him to put the pork, veal and chicken livers through the mincer. Take this home, mash it up with a fork, blend it with all the other ingredients and stir until the mixture is even and smooth. Take this back to the butcher for him to stuff into the chicken before sewing it up.

When he delivers the bird, put the calf's foot (split) and all the bones (including the chicken's) on to simmer for about 1½ hours.

Cut up the onions, carrots and mushrooms and, using a casserole that will hold the chicken comfortably, set them to colour in the butter. Place the bird on top and spoon the melted butter over it until it starts to glisten. Warm the brandy and pour it over the chicken and set it alight. This burns off the excess butter as well as improving the flavour of the sauce.

When the flames die down, cover the bird with the stock, adding a few of the chicken bones and the calf's foot. Season well, cover tightly and leave the casserole in a slow oven, Regulo 2 or 310 deg. F., for 2 hours. Uncover for the last 20 minutes.

Remove the bird from its liquor, which should be strained and left to set. After an hour or so in the fridge it should be a solid jelly. Skim every scrap of fat off the surface before slightly melting some of the jelly and returning it to the fridge. Just before it starts to gel (in about ½ hour) spoon it over the bird and leave it to set. Repeat this process once or twice. Serve with a well seasoned rice salad and dressed hearts of lettuce.

If you are serving the *ballottine* as the main course of a dinner party, precede it with a very light creamy soup made from a *purée* of fresh young peas, spinach, sprue asparagus or lettuce and follow it with an ice-cold compote of apricots spiked with Kirsch.

The whole adds up to a perfect menu for a simply prepared and easily served summer dinner party.

* *Cold Roast Duck, Wine and Orange Sauce* *

> 4½-5 lb duckling (its feet and bill should be yellow, windpipe
>   pliable, the flesh white and firm without being shiny. Avoid
>   the quick frozen birds if you can)
> juice of 1 large orange
> ¼-½ gill white wine
> ¾ gill giblet stock

Make the giblet stock with 1 pint water, ½ onion, ½ chopped
calf's foot, *bouquet garni*, 2 bacon rinds and the giblets of the bird.

Remove the duck from the refrigerator some hours before it is
to be cooked and set the giblet stock on to simmer.

Wipe the duck inside and out with a clean cloth, season it with
salt and stuff 1 tblsp of butter and either a small peeled orange or
the zest of half an orange into the cavity. Season the outside well
with salt and pepper, and smear a little butter on the breast.

Place the bird on the grid shelf of the oven preheated to Regulo 7
or 425 deg. F., and a baking pan on the shelf immediately below to
catch the fat. After 10 minutes reduce the heat to Regulo 4 or
355 deg. F. and continue cooking, allowing 15 minutes to the pound.
Baste regularly.

Halfway through the cooking pour the wine and orange juice
over the bird. It will blend with the juices from the duck to make the
basis of the jelled sauce (or gravy if the bird is to be served hot).

Mix the giblet stock, now reduced to about ¾ gill, with the juices
in the roasting pan and boil vigorously for a few minutes. Then
pour the mixture into a bowl and leave it to cool until the fat is
solid enough to skim off the surface of the sauce.

If the liquid has not jelled, bring it to the boil once again, and as
soon as it cools put it into the refrigerator to set.

Serve the bird on a bed of mint, surrounded by the chopped up
jelled sauce. Accompany with a rice salad (*see p. 181*) and a lettuce,
raddish and orange salad (*see p. 178*).

Again, a soup would be the most suitable starter as even in a
heatwave a meal is more appealing and digestible if it includes one
hot dish.

* *Green Ham* *

This Burgundian dish is pretty enough to create a minor sensation
at a supper party and yet practical enough to pull out of the fridge
for a family meal after a day by the sea.

In Burgundy, where it always appears in huge, homely white bowls, *jambon persillé* is an essential part of the Easter feast. In England I reckon it is more of a dish for salad and sunshine days.

An added refinement or variation for the second time of serving is to accompany it with a cold, soft-boiled egg (*see p. 68*).

As joints of ham are not easily come by in England, I always make do with gammon, which is an effective substitute.

2½-3 lb gammon
2 lb knuckle of veal (flesh and bone)
1 calf's foot
1 onion, 1 carrot
bouquet garni of celery, thyme and parsley
1 bottle dry white cooking wine (I use a Spanish Chablis)
1 tsp tarragon vinegar
4 tblsp chopped parsley
a few white peppercorns and a very little salt

Scrub gammon and leave it to soak for several hours, or overnight, then cover with fresh water, bring to boil slowly and let it simmer for about 50 minutes. Meanwhile, cover the cut-up knuckle and split calf's foot with the wine, add onion, carrot, peppercorns, salt and the *bouquet garni*; simmer in closed pot for the same time.

Remove gammon from its water, wash it, peel off the rind and cut it into matchbox-sized lumps. Strain the wine stock through a fine sieve lined with a piece of old (but sterile) tea towel or napkin. Pour the stock over the gammon and the calf's foot (rescued from the stock pot), cover and leave to simmer slowly until the ham is so soft it can be cut with a spoon.

Strain stock again in the same way as before and leave to cool so that the fat can be skimmed off the surface. Remove calf's foot, then using a potato masher crush the ham into pieces. Complete this breaking up process with two forks, one in each hand. Melt the fat-free stock, add vinegar, adjust seasoning if necessary, stir in chopped parsley and pour it over ham. Leave in a cold place to set.

A salad bowl, a *soufflé* dish or a pretty pottery casserole are the best vessels for this dish.

✳ *Sea Trout* ✳

During the summer months I am tempted to extravagance, not by the great salmon whose summer quality does not usually match up to its summer price, but by its more delicate relation, the sea trout.

Sea (or salmon) trout are generally sold whole. One weighing 2 lb will make a generous meal for four appreciative diners.

It responds best to the very simplest treatment — either poaching in a *court bouillon* or baking in aluminium foil.

To poach the fish make a *court bouillon* with equal quantities of white wine and water and season it with tarragon (dried will do), peppercorns, salt, parsley and a bay leaf. Bring this mixture to the boil and simmer for 10 minutes before pouring it over the fish.

Bring back to boiling point on top of the stove and then transfer it, allowing 18 minutes a pound, to a slow oven, Regulo 2 or 310 deg. F. I use a dish by Le Creuset for this, or a huge aluminium baking pan kept specially for the purpose.

If you have not got a dish big enough for the fish, wrap it securely in oiled foil. Lay it on a baking sheet and cook in a slow oven, Regulo 2 or 310 deg. F., for 45 minutes for a 2 lb fish.

Either green mayonnaise (*sauce verte*) (*see pp. 190-1*) or a slightly piquant cucumber sauce (*see p. 191*) would make worthy accompaniment to this noble fish.

Sea trout is also superb hot, of course, when its natural partner is *Hollandaise* (*see p. 185*) either plain or transmuted into a *Maltaise* with the addition of a little juice and zest of an orange.

✳ *Mackerel* ✳

Good fresh mackerel are easily recognised. Their bodies are stiff, gills red, eyes bright and colours pellucid.

One of the easiest ways of cooking them is to bake them in aluminium foil in the oven, either with or without a stuffing.

Clean the fish, but do not remove head or tail. Make an incision in the belly and stuff the cavity with seasoned butter and a *bouquet garni*. Wrap in foil and bake in the oven, preheated to Regulo 4 or 355 deg. F. for 20 minutes.

Serve cold with a sour cream and mushroom sauce. They can, of course, be eaten hot in their own juice.

✳ *Aioli Garni* ✳

This is a cross between a feast and a ritual.

The golden *beurre de Provence* (as the garlicky mayonnaise is called) is piled high in the centre of a huge dish and encircled by a miscellany of vegetables; some plain boiled — artichokes, French beans, potatoes, beetroot, baby carrots ; some raw — tomatoes, celery, peppers; and it is accompanied by hardboiled eggs, mussels, fresh or salt cod and lobster.

Even snails figure in a really grand *aioli*, but one of its charms is that it can be as grand or as simple as the occasion and the convenience of the cook dictate.

Generally, a balanced selection is made from what is seasonal, but, however it is composed, an *aioli* is virtually a complete meal in itself, needing only a little fruit to play it out and a lot of strong red wine to wash it down. At least six people are needed to do it justice and it is a simple and imaginative way of serving supper to a couple of dozen friends — given they like garlic.

Sauce for 6 people
> 6 fat cloves of garlic (about 2½ oz in all)
> 2 yolks of eggs
> 2½ gills of olive oil
> lemon juice, salt and pepper

This mayonnaise is best made in a mortar or a large heavy bowl. Crush the peeled garlic with a little salt. Add the egg yolks and stir thoroughly before beginning to drip in the oil, just as you would in the making of an ordinary mayonnaise. As the mixture starts to emulsify, the oil can be added a little faster, but if any appears unabsorbed on the surface of the sauce, stop pouring until it has been worked into the mixture.

When all the oil is used up and mayonnaise is thick and shiny like an ointment, season with lemon juice, a little more salt and freshly ground pepper.

* Fish Salad *

> ½-¾ lb halibut or turbot steak
> ½ large or 1 small lobster
> 2-3 oz fresh shrimps or prawns
> milk and water

Flavour the milk and water by adding a quartered onion, parsley, mace, salt and peppercorns, bringing it all up to boiling point and leaving it to infuse for half an hour or so before straining it over the fish. Bring gently to boiling point and poach for 10-15 minutes according to thickness. Leave to cool in the milk before removing bones and skin and breaking fish up into flakes.

Dig the boiled lobster out of the shell (if it is a big one and only half is being used, keep the rest and the coral for making into a salad with mayonnaise), and peel the shrimps. Mix both with the flaked white fish and then dress with the following sauce :

Two shallots or spring onions, a small handful of parsley, 1 level tsp of capers (all chopped very finely), a touch of mild French mustard, salt and fresh black pepper, 1 dessertsp (or a little more if you like it) of lemon juice and ½ gill of olive oil — the latter very slowly stirred into all the other ingredients.

Spoon the dressed fish into leaves pulled from the heart of a fresh cos lettuce and surround it with a rice salad (*see p. 181*).

✳ *Green Cheese Mousse* ✳

Like so many of the dishes in chapter 5, green mousse can be served as either a rich starter to a plain meal, or as the main course at a summer lunch.

It is enchantingly pretty, with the soft spring colours of the mousse contrasting with the crisp greens and reds of the filling.

> 4 oz creamy gorgonzola
> 2 egg yolks and 3 whites
> 2½ sheets gelatine
> ¾ gill fresh double cream
> 3 tblsp sour cream
> Tabasco, salt and pepper
> 2 tblsp water

If leaf gelatine is used, break it up, cover with cold water and leave to soften for 20 minutes or so. Drain it, put it in a small pan with the 2 tblsp of water and leave it to warm until dissolved : it must not boil. With crystals the procedure is the same, except they do not need preliminary soaking.

Beat the egg yolks with 2 tblsp of the cream, add dissolved gelatine and then the cheese, thoroughly mashed with sour cream. Beat until smooth.

Whip the rest of the cream and fold it in. Season with salt, pepper and Tabasco.

Put the mixture in the refrigerator to chill for 10-15 minutes before gently folding in the egg whites — whipped till they stand in peaks. Turn into an oiled ring mould (the flat type) and put into fridge to set.

Turn out by first loosening both inside and outside edges with a hot knife, and then dipping the base of the mould into hot water, *for a moment only.*

Fill the centre with an undressed mixture of cucumber, green pepper, onion, tomato (skinned and drained of juice), fennel (if you can get it), a few walnuts and a very little crisp eating apple.

Serve with pumpernickel or crispbread.

* *Boiled Beef and . . .* *

A piece of well salted silverside with or without the carrots can yield a variety of fine meals — cold with a *salsa verde* for instance (which is merely the Italian interpretation of a *vinaigrette* — though thicker and more elaborate than the French sauce), or with grated red cabbage and onion, or a very creamy horseradish sauce combined with chopped walnuts, or hot with tiny spicey horseradish dumplings.

Not all butchers put beef into brine every day, but most of them will do it to order. It is always advisable to ask your butcher just how strong the brine solution was, as a very salty joint would need some soaking before being boiled, whereas a mild pickle would be washed away in a water bath.

Being a lean cut (it is from the very top of the hind leg) and one that is not destined to be roasted away, a little silverside goes a long way — 2-2½ lb should provide ample meat for six.

Bring a pan of water to the boil and drop in the meat. As soon as the water comes back to the boil, reduce the heat until it is just at simmering point, moving but not actually bubbling.

Skim before adding a *bouquet garni* of parsley, bay leaf and thyme, an onion stuck with 3 or 4 cloves, and 2 more onions cut in half; 2 or 3 quartered carrots; a stick of celery, mace and peppercorns. Cover the pan and cook for about 2 hours.

If the meat is to make a cold meal, leave it to cool in its own liquor. Serve with a *salsa verde*, pumpernickel or rye bread and butter, and a salad of cos lettuce and watercress dressed only with olive oil.

–9–
PASTRIES AND PIES

An absurd *mystique* has grown up round the business of pastry making, as it has around the making of bread, mayonnaise and *soufflés*. In fact, as with many other minor mechanical skills, success with pastry depends on obeying a few basic rules.

·1· Keep everything (except fat) as cold as possible. This does not mean that good short pastry can only be made on a marble slab in an ambient temperature of 32 deg. Fahrenheit. What it does mean is :

> It pays to hold your hands and wrists under cold water before handling the ingredients.
>
> The liquid used for binding the paste should be as cold as possible.
>
> That all pastry is easier to roll out if it can rest in a cool place for a while, and some pastry such as *pâte sucrée* and *quiche paste* must be rested in the cool between mixing and rolling.

·2· Handle pastry as little and as lightly as possible. Unlike bread, pastry should never be kneaded or knocked around once the mixture has cohered.

• 3 • Err on the mean side with the mixing liquid — you can always put in a bit more if the paste is too crumbly, but you can't take it out. Mix in ⅔ of the amount recommended in any recipe, then, if the paste is not properly bound, add the rest of the liquid. It is difficult to give precisely accurate instructions in this matter because the absorbency of flour varies immensely.

• 4 • Always use plain flour for pastry unless a particular recipe gives explicit instructions to the contrary.

• 5 • The best tasting pastry is always made with butter.

• 6 • Pastry shrinks in the oven, so be generous with it when lining a ring or covering a pie dish. Never try to stretch it.

A FEW DEFINITIONS

For the purposes of this book only, a *pie* is any food, sweet or savoury, totally enclosed in pastry.

A *flan* is an open pastry case displaying its contents, either sweet or savoury, to the world.

A *quiche* is the same as a flan except that the filling is always savoury and that it hails from somewhere in France.

A *tart* is an expression which my sub-editor on the *Sunday Telegraph* does not allow me to use.

* Short Pastry *

These short pastries are suitable for covering all the *pies* in this chapter. The inclusion of an egg yolk in the recipe is not mere extravagance. When the proportion of fat to flour is high — as it is here — egg makes a more effective binder than water alone. Also, pastry mixed with egg has a crisper texture when it is cold.

> 8 oz plain flour
> 6 oz butter
> 1 egg yolk
> 2-3 tblsp water
> 1 dessertsp salt or castor sugar

Sift the flour and salt together, work in softened (but not melted) butter between fingers and thumbs until the mixture has the consistency of breadcrumbs. Stir 1 tblsp of the water into the beaten egg yolk and tip it into the flour. Work the mixture to a stiff dough, adding as much as is needed of the rest of the water. Leave the

paste to rest in a cool place for at least 30 minutes. *If this pastry is to be used for sweet pies, replace the salt with 1 dessertsp of castor sugar.*

Alternative short pastry which can be rolled without resting :

  8 oz plain flour
  6 oz butter
  4-5 tblsp water
  ¼ tsp salt (or 1 tblsp sugar for sweet pies)

Mix as above.

* *Pâte Sucrée* *

Sweet flan pastry should be as delicate as the soft fruit it encases and as crisp as shortbread.

The only pastry able to boast all these qualities is the French *pâte sucrée*, which is mixed with eggs instead of water and contains as much sugar as it does butter. As this makes it rather sticky to handle it is essential to allow at least an hour (or a night) for the paste to rest in a cool place between mixing and rolling. For a 7-inch flan allow :

  4 oz plain flour
  2 oz castor sugar
  2 oz soft butter
  2 small egg yolks
  grated lemon zest and salt

Sift the flour with a pinch of salt on to a board or pastry slab and make a well in the centre. Break the egg yolks into a bowl, add butter and sugar and stir until the three are roughly blended, then tip the mixture into the well in the flour and lightly but quickly draw the flour into it. Knead for a second or two before wrapping the paste in paper and putting it in the fridge to rest.

Roll out the pastry as thinly as you dare, ideally to the thickness of a halfcrown. Transfer it to the buttered flan ring and do not worry too much about the odd crack or hole, one of the virtues of this pastry is that it responds to patching.

Some flans, especially those filled with soft fruit, are best cooked blind — and filled afterwards. If that is the case, prick the bottom of the flan with a fork, line it with greaseproof paper and cover with dried beans or peas and bake in a preheated oven at Regulo 5 or 380 deg. F. for 10-12 minutes. Remove the paper, beans and the flan ring, and return the pie shell to the oven for another 5 minutes. When cold, paint the inside with white of egg.

∗ *Quiche Pastry — 1* ∗

The best behaved pastry of all.

> 8 oz plain flour
> 4 oz butter
> salt, pepper
> ½ gill water

Sift the flour with salt and pepper. Put the water and softened butter into a well, made with the fingers, in the middle of the seasoned flour and work all the ingredients together until smoothly blended. Leave the paste for at least 1 hour before rolling out.

∗ *Quiche Pastry — 2* ∗

For a 7-inch flan

> ½ lb flour
> ¼ tsp salt
> 1 small egg
> 2 or 3 tblsp water
> 4 oz softened butter

Put the egg, butter and water into a hole made in the middle of the flour which you have sifted with the salt on to a board. Let the fat absorb some of the liquid by working it with the finger tips before drawing the flour into the mixture. Work quickly and lightly, kneading the paste as little as possible. It will seem impossibly sticky when first made, but an hour or so resting in the fridge will turn the tacky mass into a manageable paste.

If it is to be baked blind, roll it out, and press it gently into a flan ring set on a baking sheet, both greased. Prick the base, cover with greaseproof paper and fill with dried beans or rice. Bake in a preheated oven at Regulo 6 or 405 deg. F. for 25 minutes, then remove the paper and beans and bake for another 5 minutes.

• • • ∗ • • •

* *Chicken Pie* *

The character of this dish is determined largely by the flavour and quality of the stock.

If inclination or the occasion demands a puff pastry, a quick frozen paste could replace the short pastry in any of the *pie* recipes which follow.

For 4-5 people

> short pastry 1 or 2
> ¾-1 lb cooked chicken (chicken boiled in the manner described on page 103 is ideal for this)
> ¾ pint stock
> ½ gill cream
> 1 generous oz butter
> 1 scant oz flour
> salt, pepper and nutmeg

Optional extras:

> sliced mushrooms tossed in butter
>
> green pepper blanched in boiling water for 5 minutes and cut in slivers.

Make the stock by reducing the liquid in which the bird was cooked or boiling up the bones and giblets of a roast bird or use a well flavoured veal stock.

Using the top of a double saucepan, melt the butter and when it is liquid, but not sizzling, slowly stir in the flour. Cook for a few moments over a gentle heat until the mixture starts to burst into bubbles like a mud geyser. Remove from the heat and slowly stir in the stock, which should be warm but not boiling. Return to the heat and bring to boiling point, then set the pan into its base (in which there is boiling water) and leave the sauce to cook for at least 30 minutes, stirring at regular intervals. Return the pan containing the sauce to the direct heat, add cream and season generously. Simmer for 2-3 more minutes, stirring all the time.

Add the chicken, and either mushrooms or peppers if used, and turn the mixture into a pie dish. Cover with pastry and bake in a preheated oven at Regulo 5 or 380 deg. F. for 40 minutes.

This is a very useful pie because it can be made the day before it is wanted and heated up.

* *Eel Pie* *

For 4 people
>   short pastry 1 or 2
>   4 small eels
>   mace and onion
>   1 gill cider, 1½ gills water
>   4 tblsp chopped shallots
>   parsley, thyme, bay, fennel
>   salt, peppercorns and lemon

Eels must be bought *alive*, but this does not mean you have to bring them still squirming into the kitchen. Get the fishmonger to behead these romantic creatures and persuade him to peel off the skin and cut them into 2-inch sections.

It is simpler to ignore the sanguine instructions given in some cookery books to fillet eels before cooking, which takes the dexterity of a *couturière*. Snip off the fins with scissors, cover the eel sections with cider and water, add lemon zest, herbs, quartered onion, salt, mace and peppercorns. Bring slowly to the boil and then simmer gently for about 15 minutes.

When the eel is tender lift it out of its liquor and as soon as it is cool enough to handle ease the meat away from the bones and lay it in a pie dish. This far the preparations can be done a day or several hours before dinner.

Season the eel with salt, pepper and lemon juice, add shallots, cover with the strained cooking liquor and top with pastry (not forgetting to make a couple of escape holes for the steam). Paint the pastry with milk and bake in a preheated oven for about 30 minutes at Regulo 6 or 405 deg. F.

* *Steak and Kidney Pie* *

English cooks argue as heatedly about the proper method of preparing steak and kidney pie as Marseillaises do about the basic ingredients of The One True *Bouillabaisse*. Even the austere *New Statesman* once gave a few of its solemn column inches to this classic controversy of the British kitchen — should the steak and kidney be cooked before or after it goes under the pastry ?

My answer to this one is *after*, unless you feel impelled to use sinewy beef from the shin.

For 4-5 people

> short pastry 1, or bought puff paste
> 1 lb rump steak
> ¼ lb veal or ox kidney
> ½ small onion, bay leaf, parsley
> salt, black pepper, dried mustard, flour
> water and a little good stock

Trim fat off beef and kidneys, and cut into walnut-sized chunks. Put seasoned flour into a robust paper bag, add meat, about ⅓ at a time, and shake up and down. Melt butter in frying pan, brown each bagful of the floured meat before transferring it to the pie dish, in which you have put a small bay leaf, and sprinkle each layer with finely chopped onion and parsley. When all the meat is browned and packed in the dish, fill the dish ¾ full of water. Cover with the pastry, decorate, make an escape hole for the steam and glaze with an egg beaten up with 1 tsp each of salt and water. Bake at Regulo 7 or 425 deg. F. for 15 minutes. Then reduce heat to Regulo 4 or 355 deg. F. for another 1½ hours. Look at the pie from time to time, and when you think it is brown enough cover it with a double thickness of dampened greaseproof paper.

✳ *Veal and Ham Pie* ✳

For 5-6 people

> short pastry 1 or 2 or bought puff paste
> 1½ lb veal — ideally from the top of the leg
> ½ lb gammon
> 1 hardboiled egg
> ¾ pint (approx.) concentrated stock which will jell, made from
>   veal trimmings and a knuckle bone
> ½ chopped onion
> mixed chopped fresh herbs, grated lemon zest, salt and black
>   pepper

Make the stock first and, if a test sample does not jell after suitable reduction, help it out with *Maggi* crystals. Cut the veal into shreds, the bacon into small pieces and the egg into slices. Season the veal and mix it with the onion, herbs and zest of lemon.

Put a thick layer of veal in the pie dish, cover with a little chopped bacon. Repeat till all the meat is used up, finishing with bacon, putting in a layer of hardboiled egg halfway through the operation. Fill up with stock, cover with pastry and bake in a preheated oven at

Regulo 5 or 380 deg. F. for 1½ hours. About halfway through the cooking, cover the pie with a piece of dampened greaseproof paper to prevent the pastry over-cooking.

When the pie is cooked, pour in a little more stock through the escape hole for the steam, using a funnel. It can be eaten hot but is, in my opinion, better cold.

* *Rabbit Pie* *

For 6-7 people

> short pastry 1 or 2 or bought puff paste
> 2 rabbits, jointed
> 1 onion, 1 leek, 1 carrot
> ¼ lb gammon
> ¼ lb mushrooms tossed in butter
> 1 hardboiled egg
> thyme, parsley and bay leaf, salt and pepper
> chestnuts (an optional extra for use in the autumn)

Soak the joints of rabbit overnight in water slightly acidulated with vinegar. Transfer the drained meat to a saucepan, just cover it with stock, or a light beer, add vegetables, herbs and seasonings. Bring slowly to the boil and simmer gently for 25 minutes.

Blanch the gammon in boiling water and cut it up into thumbnail-sized pieces. When the rabbit is cool enough to handle, remove the meat from the bone and place it, mixed with mushrooms and gammon and a slight extra seasoning of thyme, salt and pepper, in a pie dish. Add sliced egg and the chestnuts if used. Fill up with the cooking liquor.

Cover with pastry. Finish off and bake as for veal and ham pie. This is best hot, but can be eaten cold.

* *Pigeon Pie* *

For 6 people

Contrary to the suggestion given in some modern cookery books, it is essential that the birds used for this traditional English dish be young.

> short pastry 1 or 2 or bought puff paste
> 3 plump pigeons (cut in half lengthways)
> 4 oz ham trimmings
> 1 lb of buttock steak in the slice

4 oz butter
3 hardboiled eggs and 1 raw egg yolk
2 gills good jellied stock
parsley, salt and pepper (black and cayenne)

Into the breast cavity of each bird put a stuffing made of the pigeons' livers, butter, ham, parsley and seasonings all bound together with the egg yolk. Save a spoonful of the yolk to glaze the pastry.

Cut the steak into 6 even pieces and arrange them on the bottom of a large pie dish. On each slice of steak lay half a pigeon, breast downwards, and between each bird place half a hardboiled egg. Season once again, sprinkle with parsley, and pour over the stock.

Cover the dish with pastry. Gild the pastry with milk or a little salted egg yolk; make a hole through which the steam can escape.

Set the pie to cook in the centre of a preheated oven at Regulo 7 or 425 deg. F. After 15 minutes reduce the heat to Regulo 4 or 355 deg. F. and cook for another hour.

* *Tourte Bourguignonne* *

A superb variation on the theme of "veal and ham pie"

quiche pastry 2
½ lb pie veal
¼ lb each gammon and pork fat
3 tblsp cooking brandy
1 chopped onion, 1 clove garlic, salt, pepper, nutmeg, cinnamon, mace

Mince the meat and work in all the other ingredients. Leave in a cool place to mature.

Butter an 8-inch flan ring and lay it on a buttered baking sheet. Line the ring with paste and cut out the top. Paint the inside of the flan with white of egg and fill it with the meat mixture, rolled into little thumb-sized sausages.

Cover with the rest of the paste, paint with egg yolk and water mixed in equal quantities, make 4 escape holes for the steam and bake for 45-50 minutes in the centre of a preheated oven at Regulo 5 or 380 deg. F. This can be eaten either hot or cold.

* *Moules en Croûte* *

To fill a 7-inch flan you will need 2½-3 quarts of mussels.

TO OPEN THE MUSSELS
> 1 onion, 1 stick celery
> 1½ gills white wine
> parsley, thyme, bay and marjoram
> a slice of lemon peel, 6 peppercorns

FOR THE SAUCE
> ¾ oz butter, 1½ oz flour
> ¾ gill cream
> Tabasco sauce, plus 1½ gills of liquor from the mussels

Make a 7-inch flan case with *quiche* pastry 2 and bake it blind.

Clean the mussels in the usual way, scrubbing and scraping under running water followed by at least 10 minutes under the cold water tap. Throw out any with broken or gaping shells, as this means the mussel is dead. Put onion and celery (both chopped) into a large heavy pan with the wine, herbs, lemon peel and peppercorns. As soon as it boils throw in as many mussels as the pan will hold.

Cover the pan and turn up the heat. After 4-6 minutes remove all opened mussels (leaving them to cool) and add more of the uncooked ones. Repeat until all mussels are cooked. Throw away any that persistently refuse to open.

Remove the fish from the shell and strain the cooking liquor through a sieve lined with muslin.

Melt the butter in a small pan. Remove from heat and stir in the flour. Let the *roux* cook without colouring over a very low heat for a few minutes stirring all the time. Add a little over a gill of the liquor and cook gently till it thickens. Add cream and simmer for another 5 minutes. Season with salt and Tabasco. Add mussels and parsley.

Tip the filling into the flan case and return it to the oven, reduced to Regulo 4 or 355 deg. F., for 10 minutes.

* *Cream Cheese Quiche* *

> quiche pastry 1 or 2
> 3 oz each cottage cheese and gruyère
> 3 eggs, 3 spring onions
> 1 gill cream and 1 oz Parmesan
> salt, pepper, nutmeg, cayenne

Put the cottage cheese through a sieve, grate the *gruyère* and Parmesan, lightly beat the eggs and slice the onions. Stir all the ingredients together until smooth, and test for seasoning before turning the mixture into the pastry case — which may be stroked with a cut clove of garlic first. Bake in a preheated oven at Regulo 6 or 405 deg. F. for about 20 minutes.

* *Smoked Salmon Quiche* *

quiche pastry 1 or 2
6 oz smoked salmon
2 gills cream
salt and pepper
3 eggs and 1 extra yolk
1 oz butter

Beat the eggs lightly, add cream and seasoning. Pour this into the uncooked pastry shell, lay the salmon on the liquid, dot with butter and cook in a preheated oven, Regulo 6 or 405 deg. F., for about 25 minutes.

SWEET PASTRIES

* *Rhubarb Flan in the Alsatian Manner* *

The distinguishing characteristic of the flans of Alsace is the mixture of cream and egg in which the filling is set. When rhubarb is finished, the same recipe can be used with apples or cherries.

short pastry 1
½ lb young rhubarb
1 gill cream
½ gill milk
2 small egg yolks
1½ oz vanilla sugar

Whisk the eggs with the sugar and dilute them with the cream and milk. Fill the flan case with the rhubarb cut into inch-long chunks, smother the fruit with the custard mixture and bake in a preheated oven at Regulo 7 or 425 deg. F. for 25 minutes. Sprinkle the surface of the flan with castor sugar and return to the oven for another 5 minutes. Serve cold.

* *Honey Pie from the Isle of Siphnos* *

An enchanting Greek version of the customary curd sweetmeat is a honey and cheese pie from Siphnos. The Greeks would eat this between meals; for us it makes a perfect pudding to serve after a simple roast.

> short pastry 1 or pâte sucrée
> ½ lb cottage cheese
> 4 tblsp clear honey
> 2 eggs
> 2 oz castor sugar, cinnamon

Line a 7-inch flan tin with the pastry. Work the cream cheese with a fork, add sugar, warmed honey, well beaten eggs and finally a pinch of cinnamon. Pour mixture into flan case and bake in a preheated oven at Regulo 6 or 405 deg. F. for 30 minutes. Sprinkle with a little more cinnamon and leave to cool.

* *Apricot Flan* *

This is a French flan combining a professional glazed look with a country farmhouse flavour.

Soak 8-10 oz best quality dried apricots for 24 hours. Line a 7 or 8-inch flan ring with short pastry 1 or *pâte sucrée* and paint the inside with white of egg.

Pack the flan case with the soaked apricots, arranged in neat concentric circles, coming to a slight peak in the centre, and bake in a preheated oven at Regulo 5 or 380 deg. F. for 30 minutes, the last 5 minutes without the flan ring. Let the flan cool a little before giving it one coat of glaze. When completely cool coat again. Serve cold with cream.

THE GLAZE: 2 tblsp redcurrant jelly mixed with 2 tblsp of the water in which the apricots were soaked, then slowly heated until the jelly melts.

* *Fresh Strawberry Flan* *

Allow at least ¾ lb of fruit for a 7-inch flan of *pâte sucrée* baked blind. Pack strawberries tightly into the flan so that fruit comes to a slight peak at the centre, reserving 6 for the glaze.

Mash these up till smooth, add 4 tblsp gooseberry jelly and 1 dessertsp of water. Stir over gentle heat until it is dissolved. When it is quite cold, pour it over the strawberries.

* *Gooseberry Flan* *

Bake a 7-inch flan of *pâte sucre* blind.

Top and tail 1 lb small green gooseberries and poach them for about 3 minutes in a light syrup (6 oz sugar boiled for 4 minutes with 2 gills water). Drain off half the fruit and keep it whole. Gently press the rest through a sieve, but not all the way through. Spread what remains in the sieve over the bottom of the flan, and on top of this arrange the whole fruit.

Add 1 tblsp of redcurrant jelly to the *purée* that went through the sieve, boil together for a few minutes and, when it cools, pour over the fruit.

* *Fresh Peach Flan* *

Bake a 7-inch flan of *pâte sucrée* blind.

Allow 4 or 5 peaches according to size. Remove the skins by dipping them first into boiling water and then into cold. Divide the fruit into 4 or 8 slices and poach them in a light syrup (as for gooseberries above) for 3 or 4 minutes. When cool, lay them in over-lapping circles in the flan case. Boil up 1 gill of syrup with the skins and stones and add to it 2 tblsp of redcurrant jelly (dissolved over heat with 1 dessertsp of water). Cool, strain and pour over flan.

If you have enough patience to crack the stones, a few peach kernels sprinkled over the fruit will add piquancy to the flan.

* *Morello Cherry Flan* *

Line a 7-inch flan case with *pâte sucrée*, pack it tightly with stoned morello cherries, sprinkle generously with sugar and bake in a preheated oven, Regulo 5 or 380 deg. F., for 15-18 minutes.

* *Gougère* *

This is an airy éclair-like cheese cake from Burgundy, where cheese (the best of all partners for wine) is consumed in prodigious amounts.

It is excellent at either end of a meal accompanied, of course, by wine, or on its own at a wine party. Basically it is a combination of *choux* paste and cheese. It is not difficult to make, but try it out once on the family before producing it for a party. Your second effort will not taste any better than the first, but it is likely to look more professional.

> 6 oz water
> 2½ oz butter
> 1 tsp salt
> ½ tsp mustard
> 4 oz plain flour
> 3 egg yolks
> 4½ oz grated gruyère

Put the butter, water and salt into a pan and bring to the boil. As soon as it starts to bubble, pull it off the heat and shoot in all the flour at once. Stir until the mixture is smooth and comes away from the side of the pan. This takes about a minute.

Leave to cool before adding the yolks, one at a time. Stir vigorously, making sure each yolk is absorbed before adding another. When all the yolks have been stirred in and the mixture is malleable and shiny, work in 4 oz of the cheese and the mustard.

Butter a large baking sheet and draw a circle with your fingertip about 5 inches in diameter. Then, using a dessertspoon, arrange spoonfuls of the paste (like meringues) round the circle. Repeat this process, laying the next lot on top of the first, bridging the gaps between them.

Paint the paste ring with a half and half mixture of egg yolk and water, sprinkle with the remaining ½ oz cheese and put the baking sheet into a preheated oven at Regulo 7 or 425 deg. F. After 10 minutes reduce the heat to Regulo 5 or 380 deg. F. and continue cooking for another 30 minutes. Turn off the heat, open the oven door and leave the *gougère* in the warm for another 5 minutes. Eat while hot — or just cold.

* *Bilberry Pie* *

There is an abundance of local names for these tiny blue berries which look like a cross between blackcurrants and sloes. They are blaeberries in Scotland, whortleberries in Somerset (the Quantocks are covered with them) whinberries in Northumberland, and blueberries in the United States and London, England.

But whatever they are called they are worth the back-breaking task of picking, or a journey to the nearest greengrocer enterprising enough to stock them.

For a 7-inch pie plate you will need about $\frac{3}{4}$ lb of fruit, 1 heaped tsp flour, 2 oz castor sugar and short pastry 1.

Line a pie plate with two thirds of the pastry (reserving the best for the top) and paint it with egg white. Mix flour and sugar and roll the bilberries in it before turning them on to the pastry. Moisten the edges of the pastry with water, cover with the rest of the paste, press firmly together, make an escape hole for the steam, glaze with egg yolk, salt and water, and bake in a preheated oven at Regulo 7 or 435 deg. F. for 30-40 minutes. Serve cool or cold with plenty of cream.

–10–
VEGETABLES
AND SALADS

The average Englishwoman has two blind spots in the kitchen — sauces and vegetables. With vegetables, the trouble is that except when presented, for reasons of economy or conviction, as the basis of a dish ostentatiously labelled vegetarian, they tag ignominiously along in the wake of the main course unhonoured and unsung. The notion that a vegetable is a food in its own right, capable of holding its own as an individual dish or a separate course, has never caught on in England.

POTATOES

The chief victim of this attitude is the potato, which has the distinction of being both the most versatile of vegetables and also the most ill treated.

It is true that our home grown potatoes have so much to be modest about (as Winston Churchill once remarked of Clement Atlee), and it is almost impossible to buy the delicious waxy salad potatoes which are commonplace on the other side of the Channel. Most of us get the taste of the fine Continental potatoes only when something goes wrong with the home grown crop and there is a danger of shortage, but it is just because our home grown potatoes are not always all that we would wish that they need cosseting with a little of the cook's time, and a few flavourable ingredients.   Cream, butter, herbs and spices; cheese, bacon and stock — with such materials the most scarred old spuds can be turned into a delicate dish.

✳ *Mashed Potatoes* ✳

Too often these are a porridgy mass, beaten lifeless with a blunt instrument.   But with a little butter and seasoned milk and a lot of elbow grease they become a dish so creamy and light that it could be served as a separate course.

The secret of success is to dry off the cooked potatoes before pressing them rapidly through a sieve.   Then, with a wooden spoon, beat in — a little at a time — 1½ oz butter (for each pound of potatoes), followed by ½ gill of seasoned milk — also in instalments.

✳ *Plain Boiled Potatoes — Plus* ✳

Try serving these with or without skins according to your preference, accompanied by cottage cheese, sour cream, caraway seeds and slivers of green pepper or raw onion (or both).   Put all these ingredients on the table, each in a separate bowl and let everyone do his own garnishing.   Presented in this way the humble potato needs only a green salad and perhaps a glass of milk to make it into a complete meal.

✳ *Baked Potatoes* ✳

There is a universal conviction, not shared by me, that potatoes for baking must be enormous.   I find small potatoes bake just as effectively as giant ones and a great deal quicker.

A coating of salt applied before baking gives an appetising hoar-frosted glitter to the skin, as well as a crunchy texture and heightened flavour to the potato itself.   Prick the scrubbed potato in three or four places with a fork.   Place them on a cake rack near the top of an oven set at Regulo 7 or 425 deg. F. for 25-35 minutes.

If it is possible to get tired of baked potatoes served with plenty of butter and seasoning, try sour cream and chives or cream cheese and onion.

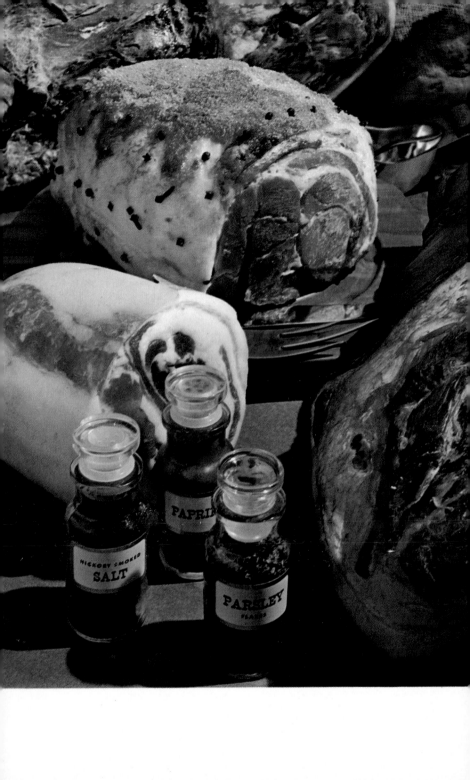

* Stuffed Baked Potatoes *

For stuffing, potatoes must be generously proportioned. Bake the potato as described above, allowing at least an hour for the cooking time. Then cut off the top and scoop out some of the flesh, being careful not to damage the skin.

The number of suitable stuffings is infinite. Try : **chopped ham and cheese; chicken livers and mushrooms** cooked together in butter; **left-over poultry,** chopped up with celery, onion and parsley and bound together with an egg; **cooked spinach** seasoned with cheese and nutmeg; **a whole egg** broken into the nearly hollow potato, topped with a little cream and grated Parmesan cheese and returned to the oven for 3 or 4 minutes; **crisply fried bacon** mixed in with the flesh of the potato.

* Potato Gulyas *

For 4-5 people

    1 lb potatoes
    ½ lb onions
    2 oz gammon or bacon
    salt, pepper, 1 tsp paprika, ¼ tsp caraway seed
    1 tblsp olive oil or pork fat
    2 gills stock or water

Using a shallow pan, brown the diced bacon in the fat. Add the chopped-up onions. When they start to colour, put in the potatoes, cut into slices the thickness of 3 halfcrowns.

Add all the seasonings, the heated stock and simmer very slowly, uncovered, for ½ hour. This should be long enough for any stock not absorbed by the potatoes to have evaporated.

* Gratin à la Dauphinoise *

This is a sturdy casserole from the Alpine region of the Dauphiné which is worthy of having a meal planned around it.

For 4-5 people

    1 lb waxy potatoes
    1 clove garlic, 1 tsp flour
    ½ pint milk, 1 gill cream
    salt, black pepper, butter

Butter, rather generously, a shallow earthenware dish. Cream the peeled clove of garlic with a little salt and spread this round the base of the dish. Using a *mandoline* slicer or the straight cutter on the grater, cut the peeled potatoes into slices the thickness of a half-crown. Arrange these in layers in the oven dish, lightly seasoning each layer.

Bring the milk to boiling point and pour it over the potatoes; finally add the cream into which the flour has been stirred. Bake in a preheated oven for 40-45 minutes at Regulo 6 or 405 deg. F.

VARIATIONS
* **1** * Sprinkle the surface generously with grated *gruyère*.
* **2** * Cover the top with sliced tomatoes and stoned olives.
* **3** * Combine 1 and 2.
* **4** * Intersperse the layers of potatoes with sliced onion.

* *Paper-Bag Potatoes* *

Potatoes take particularly kindly to cooking in aluminium foil, the modern version of paper-bag cookery, which effectively traps their elusive flavour.

Scrub 1 lb of the tiny Canary Island King Edwards, lay them on a large piece of foil spread with 1 oz of butter. Season with salt and either chopped mint or chives.

Cover with another piece of buttered foil, seal the edges and place the parcel in the centre of a preheated oven for 40 minutes at Regulo 6 or 405 deg. F.

VARIATION
Slice peeled old potatoes and spread them on an even larger piece of buttered foil. Sprinkle with 2 oz grated cheese, season with salt and pepper, and cover with ½ gill of double cream. Seal the packet tightly and bake for 45 minutes at Regulo 6 or 405 deg. F.

* *Pommes Anna* *

> **2 lb potatoes**
> **1½ oz butter**
> **salt and pepper**

Peel the potatoes and cut them into even slices no thicker than a penny piece. Leave them to soak in a large bowl of water.

Soften the butter and spread some of it on the bottom and sides of a round oven dish. Dry the potatoes and arrange them in over-lapping layers, interlarding each with butter, salt and pepper. Cover tightly with kitchen foil, and put in the centre of a fairly warm oven, Regulo 5 or 380 deg. F., for about 40-45 minutes, but take the cover off for the last 10 minutes.

Also : *potato gnocchi (see pp. 80-1).*

This chapter began with potatoes because, like bread, they are a basic food, eaten by nearly everyone every day.

The rest of the vegetables follow in alphabetical order.

Obsession with size is the occupational disease of English market gardeners — and the ruin of so many vegetables they grow: carrots the size of turnips and turnips fit only for cattle, sprouts like savoys and scarlet runners that have run away with themselves. Anyone who cares for the taste of vegetables should take the trouble and spend the money on buying the smallest possible specimens.

There are two distinguished absentees from this vegetable chapter, asparagus and artichokes. All the recipes given here are for veget-ables which can accompany meat or fish (though some of them could also stand alone), whereas these lordly vegetables are primarily a course in their own right (*see pp. 56 and 57*).

AUBERGINES

Wrinkles in an aubergine betray its age as surely as they do in a woman, so choose those with taut, well-fitting skins. If you have time, give the aubergine the salt and pressure treatment (slice, sprinkle with salt, weight down for 30-40 minutes under a plate, wash and dry). This is not essential, it concentrates their flavour.

Aubergines are normally cooked in their skins.

The simplest way of serving them is to dip the slices in seasoned flour and fry them in plenty of hot oil. Aubergines absorb a lot of oil to start with and reject some of it again as they start to crispen.

✷ Paprika Aubergines ✷

1 aubergine
1 small onion
1 heaped tsp paprika
4 tblsp oil (a little more if the aubergine seems dry)
¼ gill (⅓ of a wine glass) white wine or cider

Cut the unpeeled aubergine into ¼-inch slices. If you have time, extract the excess moisture by leaving them covered with salt for 1 hour. Heat the oil and brown aubergines gently. Cook onion finely sliced in remaining oil until golden, keeping aubergines hot in the meantime. Add paprika, and the wine or cider, and cook for 2-3 minutes, stirring well. Return aubergines to the pan, making sure they are heated through, sprinkle with parsley and serve. This is a good light dish eaten with pieces of dry toast or bread. It also goes well with lean meat, grilled or roasted, and can be reheated without spoiling. These quantities will serve 4 as a side dish or 2 as a main dish.

Also : *aubergines and yoghourt* (*see p. 62*).
aubergine casserole (*see pp. 78-9*).

BEETROOT

If you can, buy the baby ones sold in bunches, still attached to their stalks. Wash thoroughly, being careful not to bruise the skins; do not cut off the roots, and twist off the stalks gently leaving at least 1 inch of green attached to the beetroot. This is important as damaged beetroots will bleed all their colour into the cooking water. Plunge into boiling salted water and simmer for about 40 minutes. Beetroots are cooked when the skin can be rubbed off easily with the fingers.

Drain, skin and season with salt, pepper and a touch of lemon juice. Serve **hot** either tossed in butter and dressed with parsley and garlic or chives, or covered with a creamy *béchamel* sauce mixed with a little real cream. Serve **cold** with sour cream sharpened with vinegar (2 tsp for a gill of sour cream) and chives or spring onions.

BROAD BEANS

Only really worth eating while they are young enough to sit comfortably on a sixpence (though after that they can be made into a delicious *purée*). All these baby beans need is 5-8 minutes in boiling salted water and a dressing of butter or a little cream. The addition of herbs such as parsley or basil is a matter for individual preference.

BROCCOLI
(Purple sprouting)

Trim the green leaves and stalk and soak the purple heads in cold water for 1-2 hours. Plunge into boiling water, simmer 5-7 minutes, adding the salt at the last minute. Drain thoroughly. Serve **hot** with butter and lemon juice; or *sautéed* in olive oil with a touch of garlic; or dipped in batter (*see pp. 207-8*) and deep fried in olive oil; or (best of all in my opinion) dressed with an *hollandaise sauce* (*see p. 185*). Serve **cold** with olive oil and lemon juice; or *mayonnaise* (*see p. 189*).

BRUSSELS SPROUTS

Buy the smallest buds you can get (or afford), remembering the smaller the sprout the less you have to cut away. Large elderly sprouts must be stripped of all their outside leaves. If you have time, soak the sprouts in cold salted water slightly acidulated with lemon juice for 10-20 minutes. Drop them into boiling salted water, simmer for at least 5 minutes (longer, of course, if the sprouts are large), the first few minutes without a cover on the pan. Drain thoroughly and season liberally.

Serve hot dressed with plain melted butter, or with a little crisply fried bacon, or combined with glazed chestnuts.

* *Glazed Chestnuts* *

> 6 oz chestnuts
> 1 oz butter
> ½ oz sugar
> ¼ pint stock

Remove both inner and outer skins from chestnuts. This is easily done by circumscribing the sharp end of the nut with a sharp knife, covering them with cold water and bringing them to the boil. After 1 minute, remove the pan from the heat and the chestnuts — 3 at a time — from the pan. Both skins will strip off easily with the help of a knife.

Melt the butter in a heavy pan, put in the peeled chestnuts, sprinkle with sugar and let them brown all over before adding half the warm stock. Cover tightly and cook gently until most of the liquid has been absorbed by the nuts before adding the rest. Cover again and leave them to simmer until nearly all the stock has disappeared. This takes about ¾ hour. Uncover and leave the chestnuts over a gentle heat until sticky and shiny — but watch them carefully as they burn easily at this stage.

Just before serving, heat them up again (if you made them the day before) and fold them into the brussels sprouts.

CABBAGE (GREEN OR WHITE)

The French and Germans have shown us what wizardry a clever cook can work with cabbage — red, green or white, cooked or raw, braised, casseroled, stuffed or garnished. But one of my favourite recipes is English, and all of eight centuries old. It is quoted by Dorothy Hartley in her fine book, *Food in England*.

* Cabbage Seethed in Butter *

"Take a large quantity of worts — and shred them, and put butter thereto, and seethe them and serve forth — and let nothing else come nigh them." Which proves that the English haven't always maltreated their vegetables.

My own adaptation of these sensible medieval instructions is to pack the very finely shredded cabbage into a well-buttered earthenware pot, sprinkling each layer with more butter and a little seasoning, and then, having covered it tightly, stand the pot in a big pan of boiling water. Enough for 2 people can be packed into a 1 lb earthenware marmalade jar, and it takes about an hour to cook.

Cabbage prepared in this way is clean and crisp, it leaves no odour on the air of the kitchen and none of its goodness is thrown down the drain.

* Cabbage Seethed in Milk *

Another elegant way of cooking cabbage without waste or smell is to seethe it in seasoned milk.

> 1 lb finely shredded white cabbage
> 2 gills milk seasoned with 1 small onion and a chunk of celery, parsley, thyme and a bay leaf, 4 peppercorns and a shred of mace
> 2 tblsp thick cream

Drop the onion (cut in quarters), celery, herbs and spices into the milk. Bring it to the boil and then leave it to infuse while you shred the cabbage.

Strain the seasoned milk before pouring it over the cabbage, retaining only the onion, which should be chopped and put back in the pan. Cover and simmer gently for 5 to 8 minutes, just long enough for the shreds to lose their hardness, but not their crispness.

Stir in the cream, raise the heat for a moment, then serve. If you like paprika, 1 tsp sprinkled in after the cream will add subtlety to the dish.

* Cabbage Seethed in What You Will *

Cabbage takes happily to any number of liquids — cider, ham or bacon stock, water and wine, or just plain water — but the quantity should be small, the process of immersion short.

Shred the cabbage, drop it for 6 or 8 minutes into fast-boiling salted water, drain it, toss in a liberal amount of hot butter, and then season with salt, pepper, wine vinegar and sugar.

Optional additions : Chopped walnuts, slices of onion, a sharp little apple peeled and diced, shreds of celery, tiny slivers of raw pimento — any of these will tone up the texture as well as the taste of the cabbage.

* Cabbage and Sour Cream *

Cook the cabbage in the manner described above, toss it in butter, season with salt, pepper, a pinch of mustard and a sprinkling of caraway seeds. Just before serving stir in 4 tblsp (for each 1 lb) of sour cream, raise heat for a moment and serve immediately.

CABBAGE (RED)

* Spiced Red Cabbage — 1 *

$\frac{3}{4}$ lb red cabbage
1 oz butter or pork fat
1 apple and 1 onion
3 tblsp vinegar
2 tblsp brown sugar
salt, pepper

Melt fat. Chop all vegetables and add to fat. Add sugar, vinegar, seasoning, and cover tightly. Cook slowly for at least 2 hours.

* Spiced Red Cabbage — 2 *

A richer version of the previous recipe

> 1 small red cabbage (1½-2 lb) or half a big one
> rinds off 6 rashers of bacon
> 1 large cooking apple and 1 large onion
> 1 gill red wine
> 2 tblsp brown sugar
> grated zest 1 orange
> ¼ tsp each of powdered cinnamon, mace and cloves
> ¼ tsp of a mixture made by grinding white, black and Jamaica (allspice) peppercorns in a pepper mill in the proportions 3 : 1 : ¾

Using a heavy lidded pan that will go in the oven, gently cook the bacon rinds until they have delivered up most of their fat. Remove rind. Shred the cabbage finely and chop the onion and apple and tip the lot into the pan. Turn it over and over until the fat is evenly distributed and every shred of cabbage shines. Add wine, sugar, orange zest and all the spices. Salt very lightly (you can always add more at the end), stir well, cover and cook in the oven for 2½ hours at Regulo 2 or 310 deg. F. Cabbage cooked this way combines excellently with glazed chestnuts (*see pp. 165-6*).

CARROTS

Brush or scrape off the loose outer skin from 1 lb carrots. Remove the top ¼ inch from the leaf end and cut into matchsticks. Cover with water, add a pinch of bicarbonate of soda, 1 oz butter, 1 tsp sugar and a pinch of salt, and cook gently without a cover until the liquid has evaporated and the carrots are soft and slightly shiny, about 25 minutes. Serve with a sprinkling of mint.

When carrots are very young, they can be cooked whole in this manner.

CAULIFLOWER

Pick those with pure white or clear creamy curd and firm, green leaves, and beware of any where all the green has been trimmed away. It is the leaves that give away the age of the cauliflower.

Cut away the tough stalks and the outside leaves and soak for 30 minutes in salted water. Cook the cauliflower standing on its stalk in boiling water, enough to cover the green but not all the curd. As with asparagus tips, this is better steamed rather than boiled, if possible. Allow 15-20 minutes' cooking time, remembering that it should be crisp and not soggy. Drain well, break it up into sprigs and serve with : breadcrumbs fried in butter; or fried breadcrumbs with chopped hardboiled egg and parsley; or breadcrumbs fried with slivers of almond and a touch of garlic; or dressed with a *béchamel* sauce (*see p. 184*) into which you have stirred a handful of grated cheese (*gruyère* or a combination of *gruyère* and Parmesan, seasoned with nutmeg).

Also *cauliflower vinaigrette* (*p. 56*).

CHICORY

NOTE: This is the smooth pale vegetable with tightly packed leaves from Belgium, not to be confused with *endive* which, in English parlance, is a kind of tough, frilly lettuce. The confusion exists because in France these 2 vegetables swapped names.

Chicory should always be stored in the dark to prevent its leaves turning green, and if it is rubbed with a piece of lemon before being cooked it is less likely to darken in cooking. Chicory should be cooked almost dry. Put 4 heads of chicory into a small, heavy saucepan, add 1 tblsp water, $\frac{1}{2}$ tsp salt and the juice of $\frac{1}{2}$ lemon. Cover tightly and cook gently for 15-20 minutes. If they are very fat they may need 5 minutes more.

* Braised Chicory *

Clean 2 small heads for each diner. Blanch for 5 minutes in boiling salted water, drain thoroughly and transfer to oven dish. Pour over melted butter (about $\frac{3}{4}$ oz for each lb) and 2 or 3 tblsp of stock (jellied for choice). Cover and cook for about 40-45 minutes. Season before serving with salt, pepper and lemon juice.

✳ *A Gratin of Chicory* ✳

For 4 people

> 8 tightly rolled heads of chicory
> 2 oz grated gruyère and Parmesan, mixed
> béchamel sauce made with ⅔ pint milk and 1 tblsp each of butter
> and flour (*see p. 184*)
> salt, pepper, breadcrumbs
> ½ oz butter and a little stock

Bake the chicory with 2-3 tblsp of stock in a covered oven dish for 30 minutes.   Season and cover with *béchamel*.   Top with cheese and breadcrumbs and dot with butter cut into dice.   Place under a hot grill just long enough for the top to bubble and brown.

COURGETTES

The debutantes of the marrow world

Until recently one could buy these only in the Soho market or at top shops such as Harrods or Fortnums.   Now, such is their popularity, they are available in season at good greengrocers all over the country.

The true courgette is not simply a lilliputian marrow, it is the fruit of species specially developed for plucking young.

Its attraction is not only its own delicate taste, best displayed by the simplest cooking, but its ability to absorb and enhance the flavours of other foods such as meat, cheese and herbs.

But however they are prepared, courgettes taste even better if they are given the chance to drain before cooking.   This is done by cutting them in slices, sprinkling lightly with salt and then leaving them in a colander weighted down with a plate.

The simplest way of cooking is to cut them in long, very thin strips, drain as described, and then to simmer them gently in butter in a frying pan (the first 10 minutes with the lid on the pan), and sprinkled just before serving with tarragon or parsley.   Or put them in a roasting pan under a chicken that is being roasted in butter (*see p. 126*).

* Courgettes à la Crème *

For 2 people
>**1 lb courgettes**
>**1 gill single cream**
>**1 egg yolk**
>**salt, pepper and tarragon**
>**2 tblsp butter**

Cut the courgettes into long strips and drain. Melt the butter, add the strips and cover tightly. Cook on a very low heat for 15 minutes. Remove lid and cook slowly for another 5 minutes.

Mix 2 tblsp of the cream with egg yolk, and scald the rest. Pour scalded cream over the courgettes and bubble for a minute or two. Just before serving season with salt and pepper and add the egg mixture. Stir gently until the sauce thickens, but do not let the mixture come to the boil again. Sprinkle with chopped tarragon.

* Courgettes and Mint *

The perfect accompaniment to roast Kassler (*smoked loin pork*)
For two people prepare 1 lb courgettes in the usual way, only this time cut them in rounds the thickness of a halfpenny. Blanch for 5 minutes in boiling water, drain thoroughly before transferring them to a pan containing a generous ounce of foaming butter. Cook slowly for another 10 minutes, turning them over occasionally. Season with freshly ground pepper and 1 dessertsp chopped mint. Also: *Kolokythia soufflè* (*pp. 236-7*).

LEEKS

The least aggressive member of the lily family
Buy small ones if you can.

TO CLEAN : Cut off the roots and most of the green and pull away the tough outer leaves, hold the leek with the green end towards you and make 2 diagonal slashes down the length of the leek for about 2 inches. Hold under running water, flipping the cut ends with your fingers; then soak, green end downwards, in a bowl of water for as long as you can spare.

TO COOK : Allow 10-12 minutes in boiling water seasoned with salt and lemon juice. Drain very thoroughly, pressing the water out with your fingers.

Serve with: melted butter; or sour cream and paprika; or cover with grated cheddar or *gruyère* and browned under the grill; or dressed with a garlicky *vinaigrette* and popped into a moderate oven for 8-10 minutes.

MUSHROOMS

A meaty young mushroom is slightly moist to the touch and relatively heavy. Mushrooms should not be peeled and very young ones do not need de-stalking either. Hold them first under running water and then dip them into cold salted water and dry.

* Grilled Mushrooms *

Using a metal plate or fireproof *gratin* dish melt some butter and lay the mushrooms in it, gills upwards. Paint the gills with more butter and grill for 3 minutes. Turn over and grill 1 minute more.

* Sautéed Mushrooms *

Place the prepared mushrooms in a small pan with melted butter. Cover with greaseproof or foil, replace the lid, and cook on a low heat for 5 minutes. Season with salt, pepper and just a pinch of ground mace. Add 1 tblsp of cream for every 2 or 3 mushrooms (according to size), and simmer for another 3 minutes. Small mushrooms can be *sautéed* whole, large ones should be quartered.

* Stuffed and Baked Mushrooms *

For choice, use large, slightly concave mushrooms. Remove stalks and paint all over with oil and fill with the following mixture :

For each mushroom
    1 tsp chopped and lightly fried onion
    1 tsp breadcrumbs
    1 tsp chopped ham, or fried bacon
    parsley, black pepper and a touch of garlic
    1 tsp melted butter to bind

Dot with a little more butter and bake in a *gratin* dish tightly covered with foil for 20-25 minutes in a moderate oven, Regulo 4 or 355 deg. F.

ONIONS

Larger onions can be baked in the oven, like potatoes, in their skins. Nutritionally, all that is best about onions is just under the skin. Paint with oil, roll in salt and allow 1-1½ hours according to size at Regulo 6 or 405 deg. F.

To serve : Press out the solid little bulb in the centre and replace it with butter, salt and pepper.

* *Braised Onions* *

Place medium-sized, peeled onions in a buttered casserole, season with salt and pepper and half cover with boiling stock (the better flavoured the stock, the better the finished dish). Cover tightly and bake for 1 hour in a moderate oven.

PARSNIPS

To me, parsnips are tolerable only when roasted, but then they can be surprisingly good. Scrub them thoroughly, par-boil for 10-15 minutes, cut into quarters or eighths longitudinally and bake n the roasting pan under the joint.

PEAS

Garden peas, once the commonest of summer vegetables, have become a luxury in these quick-frozen days of convenient living. Few of us feel we have time for the pleasurable task of pea podding, so when we do treat ourselves to fresh, young peas, they deserve the very best treatment. Incidentally, don't fall for the fattest looking pods at the greengrocers — the seeds inside will be tough and starchy. With peas, value and quality, unfortunately, don't go together.

Allow about 3 lb unshelled peas for 4 people. Place the peas in a heavy pan with a few tender pods, a small lettuce heart, 3 or 4 spring onions, a few tablespoons of boiling water and 2 tblsp of butter. Cover tightly and cook for 20-30 minutes according to the

age of the peas.   Season with salt, pepper and sugar just before
serving.

Optional luxury — 2 tblsp boiling cream added at the same time
as the seasoning.

* *Mangetout* *

These are a type of pea cultivated for the sake of their pods which
are flat and a pale, almost translucent, green.   There is no cellulose
lining to the pod, and to prepare they need only topping and tailing
like green beans.   Also, like green beans, they can be quickly boiled
in salted water and served with butter, salt, pepper and a touch of
sugar.   Or cooked in the manner described above for garden peas.

SEAKALE

Seakale, or sea cabbage as the French call it, is the midwinter
garden's most delicate offering.   True to its name, some seakale
still grows wild, around the more remote sea coasts of this country,
and seashore conditions have been simulated in enough market
gardens to ensure a steady supply of this subtle vegetable all through
the late, lean months of the winter.

With its long, blanched leaf stalks, seakale looks like a cross
between chicory and celery and has a flavour that puts it on a par
with artichokes and asparagus.   Like these aristocrats it is at its
best plain boiled and served with a simple butter sauce.

To prepare it for the pan, cut off the black roots and clean the
stalks under a jet of water strong enough to carry away every scrap
of the sand or cinders through which the growing plants are en-
couraged to push their way to maturity.

Tie the stalks into bundles of 5 or 6, and immerse them in a large
pan of boiling salted water for 15-20 minutes.   If you have a steamer
and 50 minutes in hand, the taste and texture of the seakale will be
all the better for being allowed to cook in steam rather than water.

To my mind there is no better dressing for this delicate vegetable
than *Eliza Acton's Norfolk Sauce* (*see p. 186*).

SPINACH

Allow 2 lb for 4 people.   Young spinach needs no more than
5 or 6 minutes' cooking on a low heat in a heavy, covered saucepan

without any salt or water. It must be pressed quite dry (soggy spinach is foul), seasoned with salt, freshly ground black pepper and generously dressed with melted butter. Spinach has an almost inexhaustible capacity for absorbing butter, but I reckon that 2 oz to the pound (weighed before cooking) is a fair ration.

When the spinach is young and is to be eaten as a vegetable, there is no need to go to all the trouble of ripping out the stringy spine of the leaves, or to press it through a food mill· However, for soups or *soufflés* it is essential to reduce it to a *purée.*

* Spinach à la Crème *

For this recipe the spinach must go through the mill. Prepare it as described above, putting it through the sieve before adding the butter (which should be reduced to 1 oz) and seasoning. Then add ¾ gill of boiling cream, double for choice, and serve at once.

Spinach cooked in this way is worthy of figuring as a course on its own coming after something plain, such as grilled fish.

Spinach has other delicious uses as in *spinach soup* (*see p. 48*), *spinach soufflé* (*see p. 71*), *spinach gnocchi* (*see pp. 79-80*).

TOMATOES

The sweet subtlety of rich, firm tomatoes is a casualty of the modern demand for cheapness and uniformity. Too many of the tomatoes sold in this country today are woolly and tasteless — just how tasteless one only realises after returning from a holiday in France, Italy, Spain or Portugal where the great, irregular, yellow tinged fruit are a daily pleasure.

* Tomatoes with Cream *

Slice 1 large mild onion and soften it for a few minutes in a small pan with a tblsp of water or stock. Add 3 tblsp cream and allow it to bubble gently for a minute or two. Season with salt, pepper and a touch of sugar and basil (dried or fresh).

Meanwhile, have 4 halved tomatoes softening in a buttered dish in a gentle oven, Regulo 3 or 335 deg. F. When they are soft, cover with the cream and onion mixture and leave in the oven for another 5 minutes.

* Tomatoes in the Provençal Manner *

For each tomato allow :
1 dessertsp olive oil
1 anchovy fillet
¼ shallot or small onion
1 tblsp breadcrumbs
salt, pepper, parsley and garlic

Cut off the top of the tomato and make a hollow in the centre by scooping out the seeds and the liquid, but not the solid flesh. Chop onion finely and fry in a little olive oil with crushed garlic. Stir in all the rest of the ingredients, reserving half the breadcrumbs for the topping, and remove pan from the heat. Stuff the mixture into the tomatoes, sprinkle with remaining breadcrumbs and a little more olive oil and bake in a well-oiled shallow casserole for 25-30 minutes at Regulo 4 or 355 deg. F.

Excellent with cutlets and chops, baked and grilled fish (especially cod), or as a starter to a cold meat supper.

TURNIPS

All that I have said about size of vegetables applies to turnips twice over. The monstrous ones are fit only for cattle, or flavouring stews. The tiny ones, no bigger than mandarins, are superb.

Tiny turnips can be cooked in the same way as baby carrots (*see p. 168*), but without the sugar. They should be peeled and blanched for 5 minutes (put into cold, salted water and raised to the boil), and they will take a little longer to soften in the butter and stock. Serve with salt, black pepper, parsley and a touch of garlic.

A variation — add a little rosemary to the pan with the butter and omit the parsley from the dressing.

* Turnips in the Fisherman's Style *

1 lb young turnips
1 smallish onion
2 oz rasher of belly of pork
thyme, marjoram and parsley (all fresh if possible)
garlic and peppercorns
1 small tin of tomatoes
stock or cider

Blanch peeled turnips in boiling, salted water for 10-15 minutes, according to size. Dice the belly of pork (rind and all) and let it render in a frying pan until the fat starts to run. Add chopped onion and garlic. As soon as they take colour, add tomatoes, herbs and seasoning. Add a little stock or cider and simmer for 5 minutes. Add drained turnips and leave them to cook for 20-25 minutes. Serve with *croûtons* cut into large, thin triangles instead of the usual small squares (*see p. 38*).

• SALADS •

A plain green salad : Lettuce dressed with oil, vinegar, seasoning and fresh herbs. The best and simplest salad in the world if the lettuce is crisp and dry, the dressing delicate and the herbs fresh.

With a *mixed green salad* the spectrum widens to include a multitude of raw, green vegetables, cucumber, watercress, green peppers, endives, chicory, lamb's lettuce, batavia endive, dandelion leaves and spring onions, but not, as so many British restaurateurs seem to believe, red vegetables such as tomatoes or beetroot. Many of these salad greens can stand on their own and they all combine happily with each other.

Most green leaf vegetables are better torn than cut, and all should be dried thoroughly before being dressed.

* *Green Salad — 1* *

Lettuce, a little watercress and grated cucumber. Place a small piece of bread, spread with crushed garlic and impregnated with oil, on the bottom. Pile the clean dry leaves and the cucumber on top of the *chapon* (as this crust is called) and leave them to absorb the aroma. Just before serving dress the salad with a *vinaigrette* dressing (*see p. 181*).

Alternative dressings :

:**1** · Use double cream and white wine vinegar.
·**2** · For lettuce alone, a little crisply fried bacon (cut in dice) and its fat.
·**3** · Wine and vinegar don't mix on the palate so, if you are drinking a good or delicate wine, use lemon juice instead of vinegar in the *vinaigrette* and reduce the proportions to 6 parts oil to 1 part acid.

* *Green Salad — 2* *

An astringent, crunchy mixture that goes well with any cold meat :
watercress, cucumber and spring onions, dressed with a Roquefort
vinaigrette (see p. 181).

* *Israeli Salad* *

The reason for the name is that all the ingredients are native to Israel.
It is a splendid blend of crisp textures and softly piquant flavours,
whose secret lies in the softness of the egg yolk.

There are no hard and fast quantities, but for 2 people I allow a
small handful of torn up lettuce leaves (the crisp cos or Webb's
Wonders are best), about 1 inch of unpeeled cucumber, 2 soft-boiled
eggs (6 minutes in boiling water and then plunged straight into cold),
2 tomatoes, 4 bulbous spring onions — all roughly chopped — and
½ small green pepper finely sliced.

Season with salt and freshly ground black pepper, and dress first
with the juice of ½ lemon and finally with 2 tblsp of a fruity
olive oil. Mix thoroughly, but gently, until all the salad vegetables
have a slightly opaque coating of olive oil and egg yolk.

Made in this way a salad becomes a dish in its own right — one
to which meat, fish or cheese are pleasant but incidental accompaniments; though it is particularly good following plain grilled meat.

Served in small quantities — and with the addition of a few small
black olives — an Israeli salad also makes a perfect all-in-one *hors
d'oeuvre.*

* *Lettuce, Radish and Orange Salad* *

For 2 lettuce hearts (cos for preference) allow :

> **1 large orange (stripped of every scrap of skin and pith)**
> **9 small radishes**
> **a little chopped mint**
> **dressing made of 4 tblsp olive oil, 1 dessertsp lemon juice,**
> **½ tsp sugar, salt and pepper**

Cut the lettuce hearts into 4 with a stainless steel knife, slice the
oranges, trim the radishes and sprinkle them all with mint. Dress
the salad just before it is to be served.

Perfect with cold duck.

* Chopska Salad *

From Bulgaria, this salad is served with grills, kebabs, roasts and meat balls; or as a course on its own preceded by soup. It consists of lettuce, cucumber, tomato, onion, pepper and celery, all finely chopped, arranged in a little mound on a plate and topped with grated cheese. In winter, young leaves from the heart of a cabbage can replace the lettuce.

* Winter Salad from France *

Salads are no less essential to the balance of a meal in winter than in summer — on the contrary; in winter when the general intake of fresh fruit and vegetables, and sunshine, is reduced, a raw salad every day is even more important. Happily there are effective substitutes for lettuce, notably chicory and endives, which come into their own when lettuces pack up for the year.

This salad, a variation on the Bulgarian *Chopska* theme, is a combination of chicory, celery and *gruyère*.

Cut crisp celery and chicory into sticks about 1 inch long and leave in iced water until it is time to serve. Dress with a 5 : 1 blend of olive oil and vinegar, and season with freshly ground black pepper. Just before serving add slivers of *gruyère* (approximately 4 parts vegetables to 1 of cheese). This is capable of standing on its own as a first course.

OTHER WINTER SALADS

* Endive and Olive Salad *

Rub a crust of bread liberally with garlic, douse with 1 tblsp of olive oil and cover with 12-20 small black olives (stoned). Discard the dark outer leaves of the endive and tear the rest into the bowl on top of the olives.

Leave to stand as long as possible before seasoning with lemon juice, salt and black pepper. Mix thoroughly before serving.

This is equally good made with chicory.

VARIATION : Add sliced, skinned orange from which all pips and pith have been removed; or replace the olives with the orange.

* Orange and Onion Salad *

For each mild Spanish onion allow ½ orange and 2 sprigs of watercress. Serve with oil and vinegar dressing.

* Beetroot and Onion Salad *

Rub the bowl vigorously with garlic before adding diced beetroot. Fold in chopped onion and season with black pepper. Just before serving sprinkle with olive oil, a touch of lemon juice and parsley, or substitute cream (fresh or sour) for the oil.

* Alsatian Salad *

Finely shredded cabbage, blanched briefly in boiling water, well drained and dressed with bacon and black pepper.

* Salad of Corn off the Cob *

Use frozen corn and combine with a shredded pepper, dress with mayonnaise and garnish with a chopped hardboiled egg mixed with parsley.

* Tomato, Sweet Pepper, Apple and Onion Salad *

    4 tomatoes
    1 sweet pepper
    ½ apple
    1 small onion

Cut pepper into slivers, having removed all coarse bits and every last seed. Peel tomatoes (by plunging them in and out of boiling water and then cold water) and slice. Chop apple and finely hash the onion, combine all the vegetables, dress with olive oil, wine vinegar, salt and pepper.

TWO PARTY SALADS

* Avocado and Grapefruit *

Allow 2 avocados to 1 fresh grapefruit. Remove skin, pith and seeds from the grapefruit and slice it up. Scoop the avocado out of its skin with a stainless steel or silver spoon in thin, butter-like curls (this isn't difficult to do — it comes out that way). Dress immediately with a *vinaigrette* and garnish with parsley or any fresh herb you are lucky enough to have in the winter. This salad should not be made too long before it is needed.

* Cherry and Walnut *

Stir a 2 : 1 mixture of dark, stoned cherries and shelled walnuts into a dressing made with lightly whipped cream and cautiously seasoned with freshly ground black pepper and tarragon vinegar. Garnish with fresh tarragon. In the winter this can be made most satisfactorily with bottled cherries and a garnish of parsley.

* Rice Salad *

Boil 1 cup patna rice in a large pan of salted water till it is just cooked (12-14 minutes). Drain immediately and while it is still warm dress with tarragon vinegar, freshly ground black pepper, nutmeg and coriander, chopped herbs, slivers of spring onion and enough olive oil to make each grain shine.

When cool, fold in strips of skinned pimentos (red or green), diced cucumber, drained and chopped tomato, black olives, finely chopped celery and pine nuts — if you are near enough to Soho or a good continental grocer to buy them.

* Vinaigrette Sauce *

At its most simple, a *vinaigrette* is nothing more than olive oil, wine vinegar or lemon juice (in the proportions of about 5 : 1), salt and freshly ground black pepper. Given this basic sauce there is endless scope for embellishment — sugar, mustard, herbs, shallot, capers, Roquefort, the soft yolk of an egg, poached and sieved brains. Try :

THE ROQUEFORT addition with crisp salads (chicory, watercress, cos lettuce, apples, celery);

THE EGG MIXTURE with hot or cold boiled chicken, broccoli, plain white fish or baby broad beans;

THE COMBINATION of brains and *vinaigrette* with calf's head or hot tongue.

–11–
SAUCES

There are many ways of categorising the vastly complex subject of sauces. Mine (which is not one of the official ways) is simply to divide them into those suitable for *chefs*, and those which are practical for that woman of all work, the modern housewife.

The higher flights of the *saucier's* art, the *demi-glaces, fonds brun, fonds blanc* and all the other paraphernalia of the *cuisine rafinée* are as out of place in the private kitchen of today as the sauce bottle is on the table of a gourmet. But I suspect it is the intimidating spectre of the one and the easy availability of the other which have combined to discourage the cook from trying her hand at sauces.

Yet there are numerous attractive little sauces and dressings, quickly assembled out of simple fresh ingredients, which add immeasurably to the pleasure of eating.

The two qualities in a sauce which really matter are flavour and consistency, and flavour, which depends on the perfect freshness of the ingredients, is what matters most.

The principal ingredients of the sauces that follow are meat juices, wine, cream, butter, herbs and eggs. There is no special *mystique* attaching to any of them, not even to those which are conventionally frightening, such as *Hollandaise* and *Mayonnaise*.

· HOT SAUCES ·

✳ Béchamel Sauce ✳

If we exclude blood, there are really only 2 ways of thickening a sauce.

(*a*) With starch (which includes potato fecule, cornflour and arrow-root, as well as ordinary baker's flour).

(*b*) With egg yolk.

Béchamel is the classic example of a flour thickened sauce. It is also the most basic and adaptable of all hot sauces. And in spite of evidence to the contrary, provided by eating houses all over the country, if it is well made it has a subtle flavour of its own, as well as being a useful vehicle for other flavours.

There are 2 sources of flavour in a *Béchamel* :

(*a*) The milk which should be seasoned before the cooking starts, and

(*b*) the flour which, when heated in butter, releases a pleasant cooked sugar flavour.

Thirty minutes before starting to make your sauce, put ½ pint of milk into a pan with :

> ¼ onion
> 1 or 2 parsley stalks
> ½ bay leaf
> a little thyme (dried will do)
> a blade of mace
> 4 or 5 peppercorns

Bring the milk almost to boiling point, then take it off the heat and leave it to infuse.

Cooking the flour and butter (roux) :

Melt a *generous* ¾ oz butter in a heavy saucepan, remove from the heat and stir in a *scant* ¾ oz flour. This marginal difference in the weight of butter and flour is deliberate. It contributes to the smoothness of the finished sauce. Return the pan to a gentle heat and, stirring constantly, let the mixture bubble for a few minutes until it is the colour of Devonshire cream but no darker. Then stir in the reheated and strained milk, about ⅓ at a time, making sure that the first lot has blended before adding the rest. Bring to boiling point and cook gently for a further 3 or 4 minutes, stirring all the time. If you have time, leave the sauce to mature over a gentle heat in the top of a double boiler or in a basin over a pan of hot water. Add salt when the sauce is finished and other seasoning if necessary.

For a slightly thinner sauce, rather more milk can be added to the same quantity of *roux*. If you make the sauce some time before it is needed and want to avoid a skin forming on the top, float a thin layer of melted butter over the surface of the finished sauce. It will solidify when the sauce is cold and you can just lift it off.

* Hollandaise Sauce *

The most famous of the egg thickened sauces

This rather unorthodox recipe is adapted from one printed by that unconventional chef, Edouard de Pomiane. It is quick and almost infallible.

> **1 tsp each water and wine vinegar**
> **2 egg yolks**
> **4 oz soft butter cut in small pieces**
> **salt and pepper**

Using a wire whisk, whip water, vinegar and egg yolk together in a bowl set over hot (not boiling) water until it *starts* to thicken. Remove the bowl from the heat and add ½ the butter. Replace it and continue beating until the butter is absorbed and the mixture is creamy. Add the rest of the butter off the heat, but whip it in over the heat. After a few moments the sauce will thicken and it is ready to serve.

* Béarnaise Sauce *

This is an *Hollandaise*, scented with the essence of shallot and tarragon.

Reduce 2 shallots (or 1 tiny onion) to a hash, cover with 2 tblsp each wine vinegar and white wine, add 1 heaped tsp chopped tarragon, and season with black pepper. Boil until barely 2 tblsp of the liquid remain, then remove from heat and add 1 tsp of cold water. Turn this mixture into a bowl set over hot but not boiling water. Then proceed as for *Hollandaise*. In the winter when there is no tarragon available, use tarragon vinegar.

∗ *Hot Cucumber Sauce* ∗

> 1 small cucumber
> 1 onion
> 1 lettuce
> ½ pint chicken stock (made from giblets, bones or Maggi cube)
> 1 gill cream; 1 oz butter
> tarragon, salt and pepper

Soften peeled and shredded cucumber in the butter with onion and lettuce, also finely chopped. Add stock and seasoning and simmer till soft. Bring cream to boiling point, stir it into the sauce, add chopped tarragon and simmer for another 10 minutes. Adjust seasoning. Good with chicken, capon, salmon, sea trout and red mullet.

∗ *Eliza Acton's Norfolk Sauce* ∗

A butter sauce for asparagus, seakale or artichokes
Put 3 tblsp water into a small saucepan, and when it boils add 4 oz of fresh butter; as soon as this is quite dissolved take the saucepan from the fire, and shake it around until the sauce looks thick and smooth. It must not be allowed to boil after the butter is added.

∗ *Red Wine Sauce* ∗

For red meat

> 1 medium onion
> 1 small carrot
> 4 oz mushrooms
> 2 oz butter
> 1 tsp flour
> 1 gill red wine
> salt and pepper

Melt 1½ oz of the butter in a small, heavy pan and slowly soften the finely chopped carrot and onion until they turn golden. Don't let them brown. Add thinly sliced mushrooms and cook for another 5 minutes before adding the wine. Boil briskly for a minute, then simmer for 15 minutes until it has reduced by a third. To thicken the sauce slightly work the flour into the rest of the butter and drop this into the pan ½ tsp at a time and simmer for a minute or two more.

* *Cumberland Sauce* *

For ham, gammon and game
  4 tblsp redcurrant jelly
  4 tblsp port or marsala
  2 oranges and ½ lemon
  a scant ½ tsp mustard
  pinch each ground ginger, cayenne and coriander
  ½ shallot

Using a potato peeler, remove the zest (but not the pith) from the oranges and lemon and cut them into pin sized slivers. Slice the shallot very finely and blanch both the slivers of zest and the shallots in boiling water for 3 or 4 minutes.

Using a double saucepan, melt the redcurrant jelly, add the spices and seasoning, the orange, lemon and shallot, and finally the port or marsala. Stir until all the ingredients are completely amalgamated and then leave it to cook for another 5 minutes. Transfer the sauce to small airtight jars and store in the refrigerator.

* *Tomato Sauce for Mozzarella in Carrozza* *

Make a fresh tomato sauce by frying a chopped onion in 1 tblsp olive oil, adding chopped garlic, salt, pepper and herbs. Pour in 2 tblsp red wine (if you happen to have any) and let it bubble for a minute or two before adding a tin of tomatoes. Leave to simmer uncovered for 30 minutes.

* *Hot Tomato Sauce* *

  3 tblsp olive oil
  3 tblsp red wine
  2 onions and 1 clove garlic
  1 14 oz tin Italian tomatoes
  2 peppers (red and/or green)
  salt, pepper
  ½ small chilli (finely chopped)

Chop both onions and garlic and melt them in olive oil. Meanwhile blanch peppers in boiling water for about 3 minutes, before removing all the pith and every last pip. Slice these peppers finely and drain the tomatoes before adding both to the onion. Add wine, seasoning and chilli and simmer very slowly until the sauce is the consistency of salad cream.

* *Mild Tomato Sauce* *

Same as above, omitting the chilli.

* *Barbecue Sauce* *

For fondue bourguignonne or steak

    4 dessertsp dry mustard
    2 tblsp Barbados sugar
    ½ tsp Tabasco
    1 14 oz tin tomato juice
    2 tblsp Worcester sauce
    1 smallish onion, 1 large clove garlic
    1 stick celery
    2 gills fresh grapefruit juice
    4 tblsp wine vinegar
    1 bay leaf
    1 tblsp butter

Reduce onion and garlic to a hash and cook, till soft, in butter. Stir in dry mustard, then add all the other ingredients — the celery very finely chopped. Simmer slowly in an open pan for about 30 minutes.

• COLD SAUCES •

An extraordinary *mystique* has been allowed to grow up round the simple business of mixing a mayonnaise. Like cooking with yeast or making a *soufflé*, mayonnaise-making is something that British women seem convinced they cannot or have not the time to do. Unless, of course, they can find a machine to do it for them.

In fact, making a mayonnaise is one of the most soothing and satisfactory of kitchen tasks, and well within the capabilities of the most modest cook. And the most suitable equipment is also the simplest — a heavy bowl or mortar, a wooden spoon and a little jug with a lip. Thus armed a complete tyro might take 15 minutes, the more experienced cook between 5 and 10.

However, for women who feel that nothing they can do cannot be done better by machine, there is a small hand-powered machine called the Rapide in which mayonnaise can be whisked up in a few moments. Even husbands and children could knock up the sauce in this machine.

Another useful little gadget is a funnel with a tiny control tap at the base. It clips on to the edge of any basin and with a touch of the tap the flow of oil can be cut down to a drip or revved up to a continuous trickle.

So far, however, I have not found any electric machine which does the job more efficiently than one can do it by hand.

But whatever method is used for the making, and whatever subtle sauces are created out of the finished mayonnaise, the basic emulsion of egg yolks and oil is constant.

* *Basic Mayonnaise* *

> 2 egg yolks (or 3 if this is your first attempt — the more yolks used the simpler it is to make mayonnaise)
> ½ pint of olive oil
> ½ tsp each salt and dried mustard
> 1 dessertsp lemon juice or wine vinegar (if the sauce is to accompany fish I always use tarragon vinegar)

Break the egg yolks into the mortar, add seasonings and a few drops of lemon juice or vinegar and mix thoroughly. Now, stirring continuously, start adding the oil a drop at a time until the mixture thickens and takes on the ointment-like appearance of a mayonnaise — which it should do after about 2 or 3 minutes' dripping.

Continue adding the oil in a gentle trickle until half of it is used up. Add a touch of lemon juice before adding the rest of the oil in a slow, continuous stream. Taste for seasoning and add more lemon juice and salt if necessary.

The final result will be a thick emulsion that can be treated in a variety of ways.

· 1 · If it is not for immediate use stir in 2 tblsp of boiling water. Brutal as this sounds, it will stabilise the mixture and prevent it from separating.

· 2 · If a thinner mayonnaise is wanted (for a Russian salad, for instance), it can be diluted with either a little water, milk or cream.

· 3 · For a fluffier mixture (to accompany fish), fold in, just before serving, a whipped-up egg white or a gill of whipped cream.

· 4 · As a base for a *céleri rémoulade* it should be hotted up with some extra mustard; for a sauce to accompany a spiced beef salad add a grating of horseradish.

While there is no magic about making mayonnaise, it is worth remembering that :

Eggs should be fresh; stale yolks make poor emulsifying agents.

The more oil absorbed by the yolks the thicker the sauce will be, but no egg can hold more than 7 oz of oil, and the wise amateur won't expect it to hold more than 4 oz.

The marriage of oil and egg yolks accentuates the flavour of the oil, so it is wise to choose one that is only mildly fruity.

Contrary to much misleading advice in cookery books, the ingredients of a mayonnaise should not be too cold. The eggs should be taken from the fridge at least 1 hour before making it, and in cold weather the oil bottle should be warmed just slightly under the tap. All the talk about standing the mixing bowl on ice is necessary only if you live in the tropics. In fact, coldness is the main cause of failure in mayonnaise.

SAUCES BASED ON MAYONNAISE

* Tartare Sauce *

For ½ pint of rather mustardy mayonnaise made with 2 raw yolks and 1 hardboiled yolk add: 5 tsp of finely hashed mixture of capers, gherkins, spring onion or shallot, olives and parsley; and in addition half the white of the hardboiled egg.

* Aioli *

For ½ pint sauce, reduce 2, 3 or 4 cloves of garlic (according to your appetite for the bulb) to a paste and mix them with the yolks of 2 eggs, ½ tsp mustard, salt, pepper, and ½ tsp lemon juice. Drip in olive oil till the mixture starts to emulsify, then add the rest in a slow steady stream — using about ½ pint in all. Add a little more lemon juice halfway through and again at the end.

* Sauce Verte *

A *sauce verte* is made by working a *purée* of pot-herbs into a mayonnaise. For ½ pint mayonnaise allow about 2 oz of herbs — spinach, watercress, parsley, chives, tarragon, chervil — in the proportions of 2 parts each of spinach and watercress to 1 part of as many of the others as you can find.

Blanch the herbs in boiling water for 3 or 4 minutes, then drain, pound and push them through a sieve. Just before serving mix this *purée* into the mayonnaise.

✳ *Mustard Sauce* ✳

Two generous tablespoons *Fin de Dijon* French mustard stirred into 1 gill of sour cream.

✳ *Sour Cream and Mushroom Sauce* ✳

2 gills sour cream
$\frac{1}{2}$ lb mushrooms sautéed in butter
2 tblsp finely hashed raw onion
2 tblsp grated Parmesan
salt, pepper and garlic

Cream the garlic and stir the sour cream into it. Add all the rest of the ingredients, leaving the seasoning to the last.

✳ *Cold Cucumber Sauce* ✳

Stir a small peeled and grated cucumber (which should be ice cold) into 1 gill of lightly whipped sour cream. Season with salt, freshly ground black pepper and 1 tsp tarragon vinegar.

✳ *Horseradish Sauce* ✳

1 gill lightly whipped cream
1 tblsp freshly grated horseradish
lemon juice, salt and pepper, $\frac{1}{2}$ tsp olive oil

Fold the horseradish into the cream, add lemon juice and seasoning. Stir well before working in the olive oil drop by drop.

✳ *Salsa Verde* ✳

Reduce to a paste 1 tsp each onion and capers, a small handful of parsley, a clove of garlic, a fillet of anchovy. Stir into it slowly

2 tblsp olive oil and the juice of $\frac{1}{2}$ a small lemon. Season with salt and pepper. This sauce also makes a most interesting accompaniment to poached or fried fish, cold tongue, or the remains of a *pot au feu*.

* *Pesto* *

A Genoese sauce and the most seductive of all the sauces from the Mediterranean lands, it is a complex of mysterious tastes compounded of sweet basil, pine nuts (*pinoli*), Parmesan or Sardo cheese and garlic, worked into a dry paste and then softened to the consistency of butter with olive oil.

As I first heard of this splendid sauce through Elizabeth David, and as, except when on holiday in Italy, I have never eaten it outside my own home, I cannot do better than repeat her recipe from *Italian Food*, which is the one I always use.

> 1 large bunch of fresh basil (weighing about 2 oz after the stalks
> have been removed)
> 2 cloves garlic
> a handful of pine nuts
> a handful of grated Parmesan cheese
> just under $\frac{1}{2}$ gill olive oil
> salt

Crush the garlic, then pound it in a mortar with the basil leaves, a little salt and the pine nuts. Add the cheese. When the mixture is thick, start adding the olive oil a little at a time as you would for mayonnaise. Stir it steadily and be sure that the oil you have added is absorbed before adding any more. The finished sauce should have the consistency of soft butter. This can be made in larger quantities and stored in the fridge under a layer of olive oil.

Pesto is used for dressing *pasta*, *gnocchi*, fish, and enriching soup. The Genoese put it into *minestrone*. Their Provençal neighbours combine it with a simple vegetable soup made of potatoes, beans and tomatoes which, in honour of the special dressing, they call *pistou*.

Basil can generally be bought (from July to September) by the bunch in some big London food departments such as Harrods or Selfridges, and in Soho. Alternatively, you can grow your own.

If there is no basil there is no *pesto* sauce, but if you can't get pine nuts (though many good grocers sell them these days), an authentic version of *pesto* can be made without them.

• BUTTERS •

Savoury butters are the easiest way in the world of giving a chefly touch to simply prepared foods. Grilled fish is just grilled fish, but touch it with a little anchovy or mustard butter and you will have produced a memorable little meal. Some butters can be made in larger quantities than you require immediately and stored in the fridge.

Savoury butters are mainly used with grilled, fried or roast meat or fish, or to dress up plainly boiled vegetables.

These butters can also be used to enrich a sauce (or even a soup) at the last moment. Work the butter into the liquid briskly, just before serving, and do not allow it to come back to the boil.

* *Green (or Maître d'Hôtel) Butter* *

For 3 oz butter allow:

    1 tblsp finely chopped parsley
    1 tsp lemon juice
    salt and freshly ground pepper

Work the butter until it is creamy, then add lemon juice, parsley and seasonings. Continue working until the mixture is the consistency of a fairly stiff ointment, then transfer to a clean dish and chill. If this is to be served with steak, and everyone likes garlic, cream 1 small clove with the blade of a knife and put it into the bowl at the beginning with the butter. If garlic is added the butter will not keep.

* *Mint Butter* *

    1 oz butter
    1 tblsp finely chopped fresh mint
    salt, black pepper, lemon juice

Having creamed the butter, add the other ingredients. Use with lamb cutlets, and plainly boiled vegetables such as potatoes, carrots, peas, beetroot. This can be made and kept in the fridge in a covered jar.

* Mixed Herb Butter *

2 oz butter
2 oz fresh mixed herbs (tarragon, chervil, parsley, chives, a tiny
   bit of sorrel or spinach)
lemon juice, pepper

The herbs for this butter must be fresh. Use any 2 or 3 that you can get and make as described for *green butter*. Good with soups, fish, potatoes and carrots. This can be stored in a small jar in the fridge.

* Tarragon or Fennel Butter *

1 tblsp tarragon or fennel leaves
$\frac{3}{4}$ oz butter

Blanch the leaves in boiling water for 60-90 seconds. Run under cold water and press dry, then work into the creamed butter. Good with fish or chicken, the tarragon butter also goes well with courgettes and beans.
This can be kept in the fridge.

* Garlic Butter *

Allow $\frac{3}{4}$ oz butter for each clove of garlic. Blanch the garlic for about 2 minutes in boiling water, crush it and work it into the creamed butter.
Good with fish, mussels, *pasta* and mushrooms. This does not keep well so make it as you want it. If you are a garlic fanatic don't bother about the blanching.

* Chive Butter *

1 tsp chopped chives, 2 oz butter
$\frac{1}{4}$ tsp lemon juice

Work the chives and lemon juice into the creamed butter as described above.
Particularly good with baked potatoes and mushrooms.

* *Beurre d'Escargot or Snail Butter* *

Enough for about 3 dozen snails
> **5 oz softened butter**
> **3 large cloves garlic, a handful of parsley, 1 shallot**
> **pepper, a pinch of mixed spice, ½ coffee spoon Dijon mustard**
> **½ tsp Pernod**

It is the Pernod which gives this particular butter its distinction, but the mixture is quite satisfactory without it. Crush the peeled garlic, chop the parsley very finely and work all the ingredients together until the butter is evenly coloured throughout. Add the Pernod last of all, and continue pounding until it is all absorbed. This should be used at once as it will not stand storing.

For people who love snail butter but cannot abide (or get) snails, it can be used to dress many more homely foods, such as baked potatoes, poached eggs, mushrooms, mussels, scallops or white fish.

* *Mustard Butter* *

Work 1 dessertsp of true French mustard (*Fin de Dijon* is good) into 2 oz creamed butter. Serve with fish, particularly mackerel and herring, or with grilled gammon or hot boiled bacon. This keeps well.

* *Paprika Butter* *

> **1 oz butter**
> **1 tsp paprika, 1 shallot, softened in a little more butter**

Add the paprika to the shallot and let it cook for about 60 seconds, then work the mixture into the creamed butter.

Good with fish, veal, or for stirring into any sauce or stew containing paprika, at the last moment. This keeps well also.

* *Anchovy Butter* *

Work 3 pounded fillets of anchovy into 1 oz butter. If you are pressed, use Burgess anchovy sauce instead of the fillets. If the anchovies are very salt, soak them in milk for ½ hour before using.

Anchovy butter goes happily with many fish dishes and is surprisingly good with steak. Another good keeper.

–12–
PUDDINGS AND ICES

• PUDDINGS •

No amount of sophisticated propaganda in favour of the Continental habit of ending a meal with fruit and cheese will wean either the English man or the English young away from puddings. However, there has been something of a recession in the demand for heavy suets. Spotted dick may still be on the menu at Simpsons in the Strand, but today few adults have the appetite for solid suety sweets which defy all the rules of rational menu planning and run counter to our modern notions of healthy and beauty. The popularity of ponderous puddings dates back to the Victorian era when the German influence at the Court affected appetites and figures as well as politics. Some heavy, some rich, and all fattening, the pudding section of the contemporary cookery book reads like a guest list for Osborne — Prince Albert, Queen, Baden-Baden, Saxe-Gotha, Kaiser, Savoy, Cabinet.

Far more pleasing to our palates, and more suitable for modern menus, are the lighter confections of the 18th century, creams, fools, ices and custards.

∗ *Custard* ∗

Once upon a time, before packets were invented, a custard was a beautiful, delicate dish made simply of milk or cream and eggs. It was steamed or baked and was served as a pudding in its own right, as a garnish or a filling for tarts, or used as the basis of an ice cream.

> 3-4 egg yolks
> 2 tblsp vanilla sugar
> 1 pint milk

Bring milk and sugar slowly to a point just below boiling. Stir it carefully into the beaten egg yolks, tip the mixture into the top of a double saucepan and steam, stirring constantly, until it thickens sufficiently to coat the back of the spoon.

The basic recipe can be varied by raising the proportion of yolks to milk for a rich ice cream mixture or, as in the following recipe, replacing the milk with cream.

∗ *Crème Brûlée* ∗

In spite of a French name, *Crème Brûlée*, which was invented in the kitchens of Trinity College, is the most English of English puddings. Basically it is nothing more than a custard topped with caramel. But when the custard is a velvety mixture made with cream, and the caramel clear and crunchy, the result is an exquisite blend of flavour and a cunning contrast in textures.

> ½ pint double cream
> 3 egg yolks
> 1 oz sugar
> 1 vanilla pod

Put the cream and vanilla pod into a heavy, scrupulously clean pan, bring slowly up to boiling point, then leave it to cool. Stir from time to time to prevent a skin from forming. Meanwhile, whip the sugar and egg yolks together until white and frothy. Stir in the cooled cream and tip the mixture into the top of a double saucepan.

Stir over boiling water until the mixture just starts to coat the spoon. Turn this creamy custard into a shallow oven dish, cover and bake for ¼ hour at Regulo 2 or 310 deg. F.

When cool, cover the surface of the custard evenly with castor sugar to a thickness equivalent to that of a florin, place the dish in a baking pan half filled with cold water and put it under the very hottest grill you can contrive.

Watch this part of the operation very carefully indeed. In about 1 minute the sugar will turn from white crystals to a caramel-coloured syrup. Remove from the heat at once. When cool it will set hard and clear.

* *Hannah Glasse's Almond Custard* *

Slightly adapted

> ½ pint double cream
> 2 large eggs
> 2 oz castor sugar
>   oz crushed or milled almonds and a touch of true almond essence

Bring the cream up to scalding point in a double saucepan and, when it has cooled, pour it on to a mixture made with the egg yolks, sugar and ground almonds. Return to the double saucepan and, stirring constantly, let it cook until it coats the back of the wooden spoon. Whip the egg whites and fold them into the custard. Add essence.

Then pour the custard into little pots, set them in a pan of water, cover with foil and bake in a moderate oven, Regulo 3 or 335 deg. F., until set (about 25-30 minutes). Serve cold.

* *Syllabub* *

Syllabub is one of the oldest and most enchanting of all English puddings. Originally made by milking a cow directly into a bowl of cider or wine and skimming off the resultant foam, it derives its name from Sillery, a district in Champagne, and bub, an Elizabethan word for a bubbly drink.

This version of syllabub is generally called everlasting because it is less frothy and ephemeral than the simpler kind, and keeps well for several days. It is, in fact, best made the day before it is needed.

> ½ pint rich double cream
> ¼ pint sherry, Madeira or port
> 2-3 tblsp brandy
> 2 oz sugar
> juice and zest of 1 lemon

Mix wine, brandy and sugar in a large bowl and add the juice of the lemon. Grate the lemon rind into the cream before pouring it into the bowl with the rest. Using a wire whisk or a rotary beater, not an electric mixer, gently whip the cream and wine until it thickens enough to stand in soft peaks. Take care not to overbeat the mixture. Spoon it into wine glasses and leave in a cool place. If you have to use the refrigerator turn it down.

FOOLS

A fool should consist simply of fruit *purée* and cream. Any fruit can be used provided its flavour is not so fragile, as in apples and pears, that it would be totally extinguished by the cream that covers it, or so acid that it would sour the cream, as with lemons.

To my mind, the fruit for which fools were invented is the gooseberry.

* *Gooseberry Fool* *

Gooseberries, one of the summer pleasures of the English table, are strangely neglected by cooks on the other side of the Channel, and completely cold shouldered by the world of the *haute cuisine*.

Tante Marie simply ignores their existence. So do her cousins-under-the-apron, Madame Saint-Ange and Monsieur Pellaprat.

Escoffier, whose years at the Savoy did not allow him to ignore entirely this peculiarly English berry, did condescend to include a recipe for gooseberry fool in his book. But in lavishing his attentions upon it, inevitably he turned a simple country sweet into a courtly confection.

A perfect gooseberry fool is made as follows :

> 1 lb gooseberries
> ¼ lb sugar
> 1½ gills cream

Sprinkle the fruit with sugar and set it to soften in a moderately warm oven. About ½ hour later, when the gooseberries are cooked, press them through a food mill or nylon sieve. When the *purée* is quite cold, stir in the cream which has been slightly thickened by whipping. Serve the fool ice cold.

* *Apricot Fool* *

For an exquisite apricot fool, prepare a *purée* by putting ¼ lb dried apricots to soak for several hours, and then, without changing the water, set them to bake in a moderate oven for ½-¾ hour. Drain before putting the fruit through a sieve, or into the electric blender. Sweeten this with 1 tblsp of sugar (preferably vanilla flavoured).

Add 2 tblsp of Kirsch. Whip ½ gill of cream and fold into it the beaten white of 1 small egg. Blend the *purée* into the cream. Serve ice cold in custard glasses, decorated with chopped, blanched almonds.

SEVILLE ORANGES

Seville oranges are much neglected in modern times — except of course by the marmalade makers. Yet for many orange flavoured dishes anything a sweet orange can do a Seville can do rather better.

Orange butter for pancakes, orange whips, *soufflés* and fools, Cumberland sauce, chocolate and orange mousse, the clear orange-flavoured gravy for ducks, *Sauce Bigarade* after *bigarade*, the French for bitter orange — for all these the tart aroma of Sevilles emphasises an essential orangeness that never cloys.

* *Seville Orange Salad* *

Cut bitter oranges into horizontal slices, allowing 1-1½ for each person. Sprinkle generously with castor sugar and 1 tsp per orange of Grand Marnier. Leave to macerate for several hours.

* *Meg Dods' Orange Cream* *

> **grated zest of 2 Sevilles and the juice of 1**
> **2 egg yolks, 2 oz sugar**
> **½ pint cream**
> **1 dessertsp brandy**

Beat the eggs to a froth with the sugar, add orange zest and juice and brandy. Whip cream and fold the egg mixture into it. Leave to mature for 1 hour or so if possible.

* *Bitter Orange Tart* *

> 3 oz butter
> 3 oz soft brown sugar
> 3 eggs
> zest of 2 Seville oranges
> 1 apple
> pastry made with ½ lb flour, 6 oz butter, 4 tblsp water, 1 tblsp
>   sugar, pinch salt

Mix the pastry (*see pastry section, p. 145*) and leave it to stand
while the filling is being prepared.

To make the filling beat the butter and sugar together until soft
and fluffy, grate in the zest of orange, and then beat in very slowly
the well-whipped eggs. Add the apple, peeled and grated.

Roll out the pastry, line a flan tin and fill it with the mixture.
Bake for ½ hour in an oven preheated to Regulo 7 or 425 deg. F.
and reduced, when the tart goes in, to Regulo 5 or 380 deg. F.

* *18th-Century Flan* *

Make a *pâte sucrée* with 4 oz plain flour, 2 oz each butter and vanilla
sugar and 2 egg yolks (*see pastry section, p. 145*). Leave it to rest
for several hours or overnight. Roll it and line a 7-inch flan case.

Cream 3 oz castor sugar with 3½ oz butter until the mixture is
light and fluffy. Add, 1 at a time, the yolks of 4 eggs and the grated
zest of 1½ oranges. Beat until smooth and turn into the flan case.

Cover the surface of the flan with the grated flesh of a sweet
apple (Cox's are best) and bake it in a preheated oven for 25 minutes
at Regulo 4 or 355 deg. F. Eat either tepid or cold.

* *Junket* *

It is hard to think of anything simpler than a junket, and it is not the
sort of sweet that naturally comes to mind when planning a dinner
party. But dress it up just a little and the Plain Jane among puddings can make a most charming appearance.

> 1 pint milk
> 1 tsp rennet
> 2 tblsp brandy, whisky or rum
> 2 tblsp castor sugar

Stir spirit and sugar into the milk and gently raise it to blood heat. Remove from the stove, stir in rennet and pour it into a serving dish. Leave it in the kitchen to set, then put in the fridge.

Serve ice cold, covered with slightly whipped cream and either soft fresh fruit (such as strawberries, raspberries or blackberries), frozen blueberries, or some really luxurious jam. The Elsenham liqueur preserves (raspberry and Kirsch, cherry and brandy, strawberry and Curaçao) are ideal for this.

* Coffee Jelly *

Jelly is another nursery dish, but when it is made with coffee and served ice cold, it makes a clean and simple end to a rich meal.

For 4 people
> ¾ pint strong coffee
> ½ oz or 3 sheets of gelatine
> 2½ tblsp rum, ½ gill double cream

Make the coffee by leaving 3 oz very finely ground coffee and 1½ oz castor sugar to infuse in 1 pint boiling water for 30 minutes. Strain through a fine muslin.

Wash the leaves of gelatine, cut them up and put them to soften in a bowl of cold water while the coffee sits in a warm place on the stove infusing. Pour off the cold water before putting the softened gelatine leaves in the top of a double boiler and pouring over them a little of the strained coffee. Heat gently and stir till dissolved. Stir in the rest of the coffee and the rum. Pour the liquid jelly into a bowl, clear glass for choice, and leave it to set. Chill thoroughly and serve with ½ gill double cream floated over the surface.

* Highland Flummery *

Of the innumerable ways in which oatmeal serves the Scottish cook there is none more delicate than the distinctive crunchy finish it gives to this most elegant of puddings.

For 4 people
> 1½ gills double cream
> 3 oz (by weight) heather honey
> 3 tblsp whisky, ½ tsp lemon juice
> a sprinkling of medium oatmeal

Whip the cream until it is really stiff, then gradually stir in the honey, melted but not made hot. Add whisky and lemon juice. Serve in individual glasses, sprinkled with oatmeal which you have coloured by toasting it in the oven for a minute or two; be careful — it burns easily.

✳ *Zabaglione* ✳

This warm winey foam is my idea of instant *haute cuisine*. The ingredients, egg, sugar and fortified wine, are always to hand, and it is literally whipped up in less than 4 minutes.

For each person
>   2 egg yolks
>   1 tblsp castor sugar
>   2 tblsp wine, Marsala is best, but any other fortified wine such as sherry, port or Madeira will do

Put all the ingredients into a bowl which will fit over a saucepan containing hot (but not boiling) water and, using a rotary beater, whisk until the mixture thickens, about 3 minutes.

Serve in a wine-glass accompanied by a delicate biscuit.

SOUFFLÉS

The essential constituents of a *soufflé*, either hot or cold, are the same as for the *zabaglione*, egg, castor sugar, flavouring and AIR. The only difference is that the consistency of a *soufflé* is just slightly more stable than a *zabaglione*. In the case of a hot *soufflé* the air is trapped by cooking the egg mixture, and with a cold one by the use of gelatine.

A classic hot *soufflé* has the additional stiffening of flour, butter and milk, made into a *béchamel* sauce, which is blended with the egg yolk and flavouring before the whipped whites are incorporated. This makes the cooked *soufflé* slightly less fragile to handle, but it also makes it less fragile to eat. So for all *soufflés* to be served at the end of the meal I omit the *béchamel* base. This has the added advantage of cutting down on preparation and speeding up cooking time.

* *Apricot Soufflé* *

For 3-4 people

Make a *purée* of dried apricots as for apricot fool (*see p. 201*).

Prepare a 1½-pint *soufflé* dish by painting it with butter and sprinkling it with castor sugar. Just before it is time to cook the *soufflé*, whip up the whites of 3 eggs until they stand in peaks. Using a spatula, gently fold half the eggs into the *purée*. When it is perfectly blended, add the other half. Turn the mixture into the dish and bake in the centre of a preheated oven at Regulo 6 or 405 deg.F. for 15-20 minutes. Serve immediately.

* *Orange Liqueur Soufflé* *

> 4 eggs
> 4 tblsp castor sugar (vanilla flavoured for choice)
> grated zest of ½ orange
> 4 tblsp Grand Marnier or Cointreau
> a few fingers of sponge cake

Prepare a 1½-pint *soufflé* dish as above and cover the base with a thin layer of sponge fingers soaked in 3 tblsp of the chosen liqueur.

Separate the yolks from the whites, add sugar and grated orange and beat till pale and creamy. Add the remaining tablespoon of liqueur. Whip the whites to a snow and gently fold them into the yolks, half at a time. Pour the *soufflé* into the prepared dish and bake in a preheated oven at Regulo 7 or 425 deg. F. for 10 minutes. Serve at once.

The same recipe will make delicate lemon *soufflé* if the grated zest and juice of a large lemon replace the liqueur and the orange.

· · · * · · ·

Cold *soufflés* are extravagant with egg whites — which can be a blessing in the summer when mayonnaise and so many of the other sauces appropriate to summer foods leave a superfluity of egg whites in their wake.

* *Iced Coffee Soufflé* *

> 2 oz pulverised coffee
> ½ pint boiling water
> 1 level tblsp Barbados sugar
> 1½ sheets (or ⅔ tblsp) of gelatine
> 2 egg whites
> 1 gill double cream
> 1 dessertsp cognac or rum
> zest of orange, 3 cloves, 2 allspice berries and 1-inch stick cinnamon

Cut the gelatine into strips and leave it to soften in cold water. Put the coffee, sugar, orange zest and spices into a jug, cover with boiling water and leave it in a warm place to infuse for ½ hour.

Pour the water off the gelatine before melting it in the top of a double saucepan, or a bowl set in the top of a pan of near boiling water. Strain the coffee through fine muslin, stir it into the melted gelatine and leave the liquid to cool.

As soon as it is cold enough, put it in the fridge until it starts to set. Then stir the half-set jelly into the whipped cream. Beat the egg whites to a snow and fold them into the coffee cream. Put the *soufflé* back in the fridge and leave it to set for at least 1 hour before serving.

* *Iced Strawberry Soufflé* *

> ½ lb strawberries
> 2 oz sugar
> 2 leaves (or ¾ tblsp) of gelatine
> 4 tblsp orange juice
> 1 gill double cream
> 3 egg whites

Wash the leaves of gelatine and leave them to soften in cold water for 10 minutes or so. Drain thoroughly, cover with orange juice and heat gently in the top of a double saucepan till melted, stirring the while. Add the sugar and as soon as it has melted leave the liquid to cool just a little.

Meanwhile, press the hulled strawberries through a nylon sieve (metal wire discolours them) and stir in the melted gelatine.

Whip the cream lightly, fold in the strawberry mixture gradually and evenly and, having covered the bowl, put it to cool for about 10 minutes in the fridge, just long enough for the gelatine to start to set.

Whip the egg whites with a little salt to a fine snow and fold carefully into the half-set mixture. Turn this into individual glasses or a straight-sided *soufflé* dish, piling it as high as you can, and leave in the fridge to set for an hour or so.

The same recipe can be used with **raspberries,** use redcurrant juice to dissolve the gelatine; **blackcurrants**, use half the quantity of fruit, cook it with 3 or 4 tblsp of water and then employ the juice to dissolve the gelatine; **blackberries,** replace the orange juice with water and make a syrup of it with the sugar and, if possible, a sweet geranium leaf, and then use the syrup to dissolve the gelatine.

PANCAKES

Pancakes are for family occasions. It is not that they are difficult to make, but they do involve last-minute activity with fat and frying pan which no hostess should willingly undertake.

There is no special *mystique* about pancake making — true, the batter may be all the better for a 2-hour rest between mixing and using, but it won't be all that bad without it, while tossing is merely an amusing piece of panache.

What is essential is that the pan or griddle should be heavy, scrupulously clean and very well heated, and that the fat for cooking the pancakes be used very sparingly indeed.

The first recipe that follows is for an almost instant batter. It needs a resting period of no more than 15 minutes, and it should be cooked without any fat at all in the frying pan.

* *Quick Batter* *

2 eggs
3 oz plain flour
6 oz milk
1½ oz butter
1 tsp rum or cognac
1 tsp orange flower water (from any good chemist)
pinch of salt

Break the eggs into a hole made in the middle of the sieved flour.

Stirring briskly in a clockwise direction, gradually work in the flour until it is all absorbed by the egg. Continue beating for several minutes until the mixture is smooth and light.

In the meantime, bring the milk to the boil and drop the butter into the liquid to melt. Let the milk cool slightly before stirring it into the egg and flour paste.

Add the salt and flavourings before giving the batter a few turns with a rotary beater and leaving it to stand for 15 minutes.

Heat up the pan until the reflected heat hits the palm of your hand held 6 inches above its base. Pour in a very small quantity (not more than 2 tblsp) of the pancake batter. After a few seconds it will be ready to be flipped over with a spatula, and in less time than it took to type this sentence the pancake will be ready.

* *Jacques* *

Sugar and lemon juice, the traditional English dressing for pancakes, are hard to beat for simplicity and freshness. A little less conventional, though equally clean tasting, are the apple pancakes from the Périgord, known as *Jacques*.

Pour a little batter into the pan, place on it a few thin slices of apple sprinkled with sugar and lemon juice or soaked in a tot of brandy or calvados, and immediately cover the apple with a little more batter.

Jacques, which are a sort of apple sandwich made with batter, take rather longer to cook than do conventional pancakes, and they are slightly more tricky to turn over.

* *Walnut Pancakes* *

> 1 oz each butter and sugar
> 3 oz ground-up walnuts
> ½ beaten white of 1 egg
> 2 tsp of orange juice or orange liqueur

Beat the butter and sugar into a smooth cream, and work in the walnuts, egg white and flavouring.

Fill the pancake with this mixture and then garnish the filled pancake with a glaze made of 1 tblsp of butter, 2 of sugar and 3 of orange juice simmered together for 5-10 minutes until the mixture is syrupy.

* *Blackberry Pancakes* *

For these a slightly more airy batter is best, though if you are in a hurry use the quick batter recipe, folding an egg white into the mixture just before adding the blackberries.

> 6 oz blackberries
> 3 oz flour and pinch of salt
> 1 egg
> 1 gill milk and $\frac{1}{2}$ gill water
> 1 dessertsp each olive oil and sugar
> butter for cooking

Make a smooth batter with the flour, egg yolk and liquid. Add the salt, sugar and olive oil, beat for a few minutes and then leave it to mature until you are ready to use it. Just before making the pancakes, whip the white of an egg up to a snow, fold it into the batter before stirring in the blackberries.

Melt a scrap of butter in a heavy omelette pan or a griddle, and spoon into it just enough of the pancake mixture to cover the bottom. As soon as the base of the pancake is set flip it over and allow the other side about a minute to set. Serve with plenty of cold whipped cream.

* *Rhubarb Clafoutis* *

A robust, family pudding made on the same principle as the blackberry pancakes above.

> $\frac{1}{2}$-$\frac{3}{4}$ lb rhubarb
> 3 eggs plus 1 extra white
> 1$\frac{1}{2}$ tblsp flour
> $\frac{1}{2}$ tsp ground ginger
> $\frac{1}{2}$ gill cream and 2 tblsp milk
> 4 tblsp sugar
> Kirsch (optional)

Whip up the eggs, sift in the ginger and flour and whisk well before folding in half the sugar. Add the cream and milk and whisk once again. If you can afford the time, leave this batter to stand for a while.

Cut the washed rhubarb into tiny chunks, sprinkle it with the rest of the sugar and the Kirsch and cover it with the batter. Bake in the centre of a hottish oven, Regulo 6 or 405 deg. F., for 20-25 minutes. Serve either cold or lukewarm with plenty of cream.

* Boiled Gooseberry Pudding *

Line a pint-sized pudding basin with a thin suet paste, made with ½ lb flour, 3 oz freshly grated suet, 1 gill water, 1 tsp each sugar and baking powder, ½ tsp salt and a pinch of cinnamon.

Fill the bowl with about ¾ lb topped and tailed gooseberries, interlarding each layer of fruit with a generous sprinkling of Demerara sugar.

Seal the fruit in with a lid of suet paste, pressing it tightly on to the lining crust so that no juice can escape. Tie down securely with kitchen foil and put the pudding to steam for 2½ hours. Serve hot with lots of fresh cream.

* Bread and Butter Pudding *

Fill a shallow ovenproof dish with thin slices of white buttered bread, sandwiching each layer with brown sugar and raisins. Barely cover with an egg custard (*see p. 198*) flavoured with cinnamon. This is made by infusing 1 inch of cinnamon for 1 hour in the milk before mixing it with the egg yolk. Bake in a moderate oven, Regulo 3 or 335 deg. F. Serve cold with cream.

* Strawberry Shortcake *

As made by my mother every 23rd June to celebrate the birthday of the only member of our family with the sense to be born in the summer.

Make a sponge with 3 eggs, 4 oz sugar, 3 oz plain flour and grated zest of 2 oranges. Beat eggs and sugar until the mixture is nearly white and trebled in volume. Fold in flour, half at a time, and zest of orange very gently. Tip the mixture into buttered and floured baking tin (3 inches by 7 inches) and bake in preheated oven at Regulo 4 or 355 deg. F. for 30-35 minutes.

When cold cut it in half horizontally and spread the centre with a lavish amount of mashed strawberries (cheap berries will do for this). Sprinkle with sugar, put the two halves together again and cover the top with lightly whipped cream. Set as many halved strawberries into this as the cream will hold.

* Grapes under the Grill *

The sweet and cheap little sultana grapes from Cyprus, with their soft skins and seedless flesh, are best for this dish, which is an enchanting pudding distantly related to baked Alaska. Grapes do duty for the ice cream, and cold whipped cream takes the place of meringue.

For 4 people
>**1 lb grapes, 1½ gills cream**
>**2 or 3 tblsp soft Barbados sugar**

Chill grapes and cream ice cold. Strip the grapes from their stalks and pile them into an oven-glass casserole. Whip the cream and spread it over the fruit, and leave the casserole in the fridge until you are ready to serve it. Sprinkle with sugar and put the pudding under a hot grill just long enough for the sugar to bubble.

* Poached Quinces or Pears *

Quinces are worth searching out, not only for the bewitching savour released in cooking but also for the unearthly richness of the scent with which they fill the kitchen. If you cannot get them, use fairly hard Conference pears.

Peel the fruit but do not remove the core. Stand them upright in an oven dish containing ½ gill of water. Pour 1 tblsp of mild honey over each quince or pear, season with the merest touch of clove — it is so easy to overdo this spice. Cover tightly and cook in a very slow oven for about 2 hours. Test with a skewer. Chill and serve with whipped cream flavoured with grated orange.

* Orange Orientale *

Mario of the "Caprice" gave me this rather grand recipe which I have slightly adapted to suit the exigencies of the private kitchen.

It is a glamorous confection consisting of the whole fruit, skinned, and marinated in syrup, crowned with its own candied peel.

Remove the zest and cut it into match-like strips. Blanch these in boiling water for 8 minutes and then transfer them to a syrup made by boiling 6 oz sugar in 1 pint of water for 15 minutes. Simmer the "marmalade" slowly about 15 minutes until soft and translucent.

Strain off the peel, continue boiling the syrup for another 5 minutes, add the juice of an orange. Pour this over the stripped oranges and leave them to marinate for 24 hours. Serve ice cold topped with the "marmalade" and surrounded with syrup.

• WATER ICES •

The sweet iced flavour of fruit is a clean and refreshing *finale* to a rich meal. Water ices are cheap and easy to make, provided you have a refrigerator. The crystals which are inevitable in any ice made in the freezing tray of a refrigerator (where there are no paddles to agitate the mixture) are part of the crunchy character of a water ice, and not undesirable intruders as they are in cream ices.

* Strawberry Water Ice *

For 6 people

1½ lb strawberries — yielding approximately 1½ cups purée
4 oz sugar, including 6 lumps rubbed over the skin of an orange until they have absorbed the oils
¾ gill water
juice of ½ an orange
2 tsp lemon juice

Simmer the sugar and water together for about 8 minutes. Leave this to cool while pressing the hulled strawberries through a nylon sieve. Stir the syrup into the *purée*, add the orange and lemon juice and pour the mixture into the freezing tray of the fridge. Cover with foil and leave the ice to freeze for about 3 hours at normal refrigerator temperature.

* Blackberry Water Ice *

1½ lb blackberries — yielding approximately ½ pint of juice
6 oz granulated sugar
1½ gills water

Make a syrup by boiling the sugar and water together for 20 minutes in an open pan. Let the sugar dissolve before bringing the liquid to boiling point. Meanwhile, press the berries through a nylon sieve. When the syrup is cool, stir the *purée* into it. Freeze, covered with foil, in the ice tray of the refrigerator, set at 7. The freezing will take 2 to 2½ hours and the mixture should be taken out and whipped with a rotary beater 3 or 4 times during that period.

* *Orange Water Ice* *

Orange water ice can, of course, be served in a glass like the other water ices; on the other hand, nature provided oranges with such attractive packages that it seems a pity not to exploit them.

For 6 people
> 6 Jaffa oranges
> 6 oz sugar
> ½ pint water
> ½ pint orange juice from the pulp of the oranges
> 1 lemon
> 1 extra orange
> 6 dessertsp Cointreau
> 3 dessertsp whipped egg white

Using a potato peeler, pare the outer skin off the lemon and the extra orange. Put this, with the sugar, in a pan with the water and heat slowly till the sugar dissolves. Then boil briskly for 6 or 7 minutes. Meanwhile, cut the top off the oranges and scoop out the flesh without rupturing the skin. Crush the pulp for juice and squeeze the juice from the lemon. When the syrup is cool, mix it with the juices and put it to freeze in the ice tray of the fridge, set at its coldest. When the mixture is mushy but not solid whip it round with a fork and fold in the frothed egg white. Continue freezing until stiff. Meanwhile, put 1 dessertsp of Cointreau into each orange and swill it round. Leave the oranges in the fridge till the ice is stiff, then fill them, replace the lid and stand them in the ice compartment until it is time to serve them.

· ICE CREAMS ·

"Ices", Escoffier proclaimed in a moment of creative flamboyance, "are the consummation of all that is delicate and good."

I am not going to pretend that home-made ices are any cheaper than the bought variety — how could they be ? Double cream, whole eggs, fresh fruit and castor sugar cost so much more than dried skim milk, cornflour and fruit flavourings.

But when it comes to comparing the taste, no one who can tell rayon from silk, rabbit from chicken or marge from butter could fail to prefer the sweet summer flavour of ices made in the kitchen to the professional job made in the factory.

But Escoffier knew of only one way to make an ice — in a freezing machine with perpetually rotating paddles. And half a century later there is still no other way of making a perfect ice.

There is, however, for those prepared to spend about £10, a painless, space-age version of the old Victorian, muscle-powered machine. It is a French device known as the "Arpi", which is electrically driven and fits into the freezing compartment of most modern fridges. The Arpi is imported by Clarbat Ltd. (302 Barrington Road, S.W.9).

What distinguishes it from its predecessors is the automatic cutout. As soon as the ice-cream mixture is properly congealed, the paddles stop turning, which means you can set it going in the morning, and forget it.

What's more, it is not necessary to turn the fridge to its coldest setting, which damages salad vegetables and does no good to the milk, as you have to when making ice cream in the ice tray.

However, the nostalgics and the fridge-less among us, may be glad to know that this modern piece of machinery has not completely superseded the good old wooden ice bucket. These can still be bought — new as well as secondhand — so can the ice and freezing salt on which they depend.

(Ice buckets — new from Staines Kitchen shop, 122 Victoria Street, S.W.1, Habitat, 77 Fulham Road, S.W.3, and Harrods.) .

There are still fishmongers to be found who are willing and able to supply ice, also firms who specialise in delivering ice. It usually comes by the hundredweight block or as a large bag of cubes. The same firms generally sell freezing salt as well.

All the recipes that follow can be made to perfection in a mechanical freezer — either manual or electric. They can also be made satisfactorily in the ice tray of a refrigerator, provided they are turned out and whipped with a rotary beater several times during the freezing to prevent ice crystals from forming.

The perfect home-made ice, and also the simplest, is merely a mixture of sweet whipped cream and fruit.

I favour rather fruity ices, and find that the sieved *purée* from a pound of fruit (about $\frac{1}{2}$ pint) needs only a gill of cream to make an exquisite ice cream, though the more powerfully flavoured fruits such as blackcurrants will need a little more.

A rather less rich, and cheaper, ice cream can be made with a 50-50 mixture of egg custard and cream. For each gill of cream used allow 2 egg yolks and 1 gill of milk for the custard. Make the custard in a double boiler and when it is cold fold it into the cream whipped with 2 oz sugar.

The amount of sweetening must vary with the fruit used and the taste of the individual, but it should be remembered that the sweeter the mixture, the longer it will take to freeze.

When making ice cream in the ice tray always set the thermostat at its coldest.

* *Raspberry and Redcurrant Ice Cream* *

    1 lb raspberries
    2 oz redcurrants
    2 oz sugar
    1 tblsp water
    1 gill cream

Make a syrup of the redcurrants, sugar and water by boiling them together for 6 or 7 minutes. Press the raspberries through a nylon sieve and dilute with the syrup. Whip the cream and fold into it the redcurrant and raspberry pulp. Turn into freezing tray and freeze for 2 hours.

* *Brown Bread Ice* *

Once this ice was a Winchester College treat and later a speciality of Gunters

For 6 people
    $\frac{1}{2}$ pint double cream
    3 oz good wholemeal bread without crusts
    1 oz castor sugar
    3 tblsp honey
    1 tblsp orange or lemon juice

Cut the bread into slices and set it to dry in a slow oven for about 20 minutes. Meanwhile set the cream, lightly whipped with the castor sugar, into the freezing compartment of the fridge and leave it there for $\frac{1}{2}$ hour.

When the bread is beginning to harden, but before it turns into a rusk, reduce it to crumbs using either a grater or an electric blender. Melt the honey, add the juice to it and pour the mixture over the crumbs.

Amalgamate the semi-stiff cream with the sweetened crumbs, stirring thoroughly to be sure that it is evenly mixed. Return to the fridge and freeze for another 2 hours. Being low in water content there is no need to turn this out and whip.

* Blackberry Ice Cream *

>   1 lb blackberries
>   6 oz sugar
>   1 gill water
>   1½ gills cream
>   a sweet geranium leaf (not the common variety)

Boil sugar and water with the geranium leaf for 7 or 8 minutes. Leave to cool, then stir this syrup into the sieved blackberries. Slightly whip the cream before folding the blackberry mixture into it. Freeze in the usual way.

* Orange Juice Ice Cream *

Scald 1½ gills of single cream, stir in 4 oz castor sugar and leave it to cool. Add 1½ gills double cream, and freeze until slushy. Add 1½ cups of fresh (not tinned) orange juice, 1 tblsp lemon juice, and return to freezer.

* Apricot Ice Cream *

>   just over 1 lb fresh apricots
>   3 oz castor sugar
>   1½ gills double cream
>   1 tblsp lemon juice
>   1 tblsp Cointreau
>   almonds

Wash apricots, sprinkle with sugar and bake without water in a covered casserole in a gentle oven until soft. Work them through a sieve. Add lemon juice and liqueur. Whip the cream lightly and fold in the apricot *purée*. Freeze in usual way. When the mixture reaches the slushy stage, stir in a few blanched sliced almonds.

* *Praline Ice Cream* *

Slightly bruise a vanilla pod before putting it into ½ pint of homo-genised milk. Heat the milk, but do not boil. Leave to infuse for a while, then, having removed the pod, make a custard with the milk using 3 egg yolks, and sweeten it with 3 oz of castor sugar.

As soon as the custard is cool, fold it into ½ pint lightly whipped double cream. Freeze in the usual way. Shortly before serving, fold in 1½ oz crushed praline.

PRALINE : Using a heavy pan, heat equal quantities of unblanched almonds and castor sugar very slowly without stirring. As soon as they start to colour stir gently until the mixture turns a toffee brown, then turn immediately into a flat oiled tin. When cold, crush into a powder and store in an airtight jar.

−13−
PRESERVES

• JAMS and JELLIES •

Although, it seems, there is no time or appetite these days for that charming ceremonial, afternoon tea, this is no reason for denying oneself the satisfaction of building up a little cache of jams and jellies made from soft, summer fruits. After all, there are more ways of using an exquisite preserve than spreading it over bread and butter — for instance filling tiny tartlets made of a rich, short crust, or accompanying something cool and creamy such as fresh cream cheese, a plain homemade ice or whipped cream lightened with a beaten egg white. Soft fruit jams, ideal for these purposes, lend themselves especially well to being made in small quantities and going into small pots.

For small batches I find that the ideal pan is the bottom half of the pressure cooker. Being made of really heavy cast aluminium and having a base that is nearly half as thick again as its sides, it is ideal for a jam pan. But for jam making on a family scale a proper preserving pan is essential. It is also essential that it be made of heavy gauge aluminium or copper — the latter, of course, being unlined.   In this instance copper is likely to be the better buy. Being a better conductor of heat than aluminium, a copper pan can get away with being slightly lighter gauge, and therefore cheaper.

✳ *Dark Cherry Jam* ✳

> 2 lb good black cherries
> 2 gills redcurrant juice
> 2 lb preserving sugar

Wash and stone the cherries.   Melt the sugar in redcurrant juice and when it starts to froth, drop in the fruit.   Cook, stirring all the time, for 8-10 minutes, removing any scum that rises.

Lift out the cherries with a skimmer, and carry on boiling the juice for another 10 minutes or so until it starts to thicken.   Replace cherries and repeat the process, then turn the jam into a large basin to get quite cold before putting it into pots.

✳ *Fresh Raspberry Jam* ✳

> 1 lb each raspberries and preserving sugar

Tip the fruit and sugar into an earthenware casserole, and set them in a very slow oven, Regulo ½ or 265 deg. F., until the juice runs and the sugar starts to melt — 30-40 minutes.   Then turn the mixture into a preserving pan, and boil for just 2 minutes.   Pot at once.

✳ *Mulberry Jam* ✳

> 2 lb mulberries
> ½ pint water
> 1½ lb sugar
> 1 lemon (the grated zest and the juice)

Wash the mulberries, cover them with the water and simmer until soft.   Add the sugar and the zest and juice of the lemon.   When the sugar has melted, boil until setting point is reached, when a little put in a saucer sets when cold.

* *Quince Jelly* *

Of all confections made from quinces, nothing can surpass its simple, fragrant jelly.

Wipe the quinces thoroughly to remove the fur and slice them up without peeling or removing the cores. Cover them with cold **wate**r, allowing about $1\frac{1}{2}$ pints to each pound of fruit. Cook over a slow heat until the fruit is really soft, but not pulped. Make a jelly-bag out of a double thickness of muslin or, failing this, out of an old tea towel that is free of holes and absolutely clean. Tie the corners securely to the legs of an upturned chair so that the material between the legs is loose but not in folds. Place a large bowl under the bag and pour through this a kettleful of boiling water to clean and sterilise the material. Empty the bowl and replace it under the bag.

When the fruit is cooked transfer it gently while still hot from the preserving pan to the bag. Leave the liquid to the last and pour it slowly over the fruit in the bag, leave to drain all night. Hurrying this by squeezing the bag or pressing down the fruit usually clouds the jelly.

Next day measure the strained juice and boil it up in the pan, adding $\frac{3}{4}$ lb of sugar for every pint. The flavour of quince jelly may be varied by the addition of a few strips of the zest of an orange. Such additions should be removed before potting the jelly. A knob of butter will help inhibit formation of scum. Heat gently until the sugar dissolves, then boil rapidly until setting point is reached.

After 10 minutes of boiling, test for setting by putting a few drops of jelly on to a cold saucer and leaving it in the refrigerator. When a sample sets remove the pan from the heat. Pot immediately in small warmed jars, covering the surface with wax discs.

* *Cotignac* *

A delicious by-product of quince jelly is *cotignac*, a French version of our familiar quince cheese and a sweetmeat on which Rebecca West said she was raised.

Cotignac can, of course, be made with the whole quinces, but I see no reason not to have a few jewel-like jars of jelly as well — in which case the *cotignac* is made with the pulp that remains in the jelly bag after the juice has dripped away.

Put the pulp through a sieve, and for each pound add 1 lb sugar and the juice and zest of $\frac{1}{2}$ a lemon and $\frac{1}{2}$ an orange. Cook in a thick pan, stirring constantly until the mixture is thick and stiff.

Then spread it out in a shallow baking tin and leave it to dry in a cool oven for 1 hour. Store in airtight tin or jar.

Cotignac can be eaten on its own after dinner, or served as a pudding with cream or cream cheese.

* Five-Fruit Jelly *

> ½ lb each strawberries, raspberries, cherries, redcurrants, gooseberries
> 3 gills water
> preserving sugar

Wash the fruit and put it in the preserving pan without bothering to remove hulls, stalks or stones. Cover with water, and simmer gently for 30-40 minutes until all the fruits are soft. Turn it into a scalded jelly bag made of calico or at least 2 layers of muslin and leave to drip until all the juice has drained away.

Return juice to clean pan, bring up to boiling point, and after 2 or 3 minutes add 1 lb of sugar for each pint of juice. As soon as the sugar melts, boil fast until setting point is reached, 10-15 minutes. Pot immediately.

* Medlar Jelly *

Medlars are normally a November fruit, but the medlar season can stretch into December. They seldom appear in the shops, but there are a surprising number of medlar trees to be found by the observant, even in cities.

Medlar jelly has an astringent quality that makes it go as well with meat as it does with scones and butter or cream cheese.

Cover the washed fruit with water (1-1½ pints a pound) and leave it to simmer slowly in an open pan for 2-3 hours. When the fruit is really soft and mushy, transfer it to a jelly bag and leave it to drip overnight.

Next day measure the juice and for each pint add ¾ lb sugar and 1 tblsp of lemon juice. Heat slowly, not allowing the liquid to boil, until all the sugar is dissolved. Then boil rapidly until setting point is reached — 8-10 minutes if the jelly is made in small quantities.

Skim before pouring the jelly into small warm jars and seal when cold.

✳ *Gooseberry Jelly* ✳

There seems to be a special fragrant affinity between the gooseberry and the flower of the elder tree. A few heads of elder blossom, drawn through a panful of gooseberry jelly (or jam) just before it is ready to pot, impart to the conserve a subtle flavour of muscat grapes.

> 1 lb hard green gooseberries
> 1 gill water
> 2 heads of elder blossom, picked just before they reach full bloom, wrapped loosely in a piece of sterilised muslin

Add the water to the gooseberries (there is no need to top and tail them) and leave them to simmer for about ½ hour, by which time the fruit should be soft and the juice flowing freely. Transfer the hot pulp to a large piece of muslin (rigged up in the form of a jelly bag) and leave it to drip for several hours. A little very light pressure on the fruit at this stage will help to express the juice without clouding it.

Measure the juice, return it to the pan and stir in 1 lb warmed sugar for each pint of liquid. Let the sugar dissolve completely before allowing the jelly to come to the boil, then let it bubble briskly for 15-20 minutes until setting point is reached.

As soon as the jelly is ready, take the pan off the heat and drop in the little bag of elder flowers. In less than 5 minutes the jelly, having absorbed the delicate muscat flavour of the flowers, will be ready to pot.

MARMALADE

Breakfast, as A. P. Herbert once remarked, is a critical period in matrimony. It is the most fragile time of day, when surprises of all sorts are unwelcome, when both conversation and comestibles should be comfortably predictable. And the one thing every English man expects to see on his breakfast table is a jar of marmalade. Jam just won't do. On the other hand, off-beat marmalades will, and the Seville season is so short that more often than not they have to.

But whatever the basis of the marmalade, the basic rules are the same :

·**1**· Always wash citrus fruit thoroughly before starting as these days the skins may be contaminated with chemicals from sprays and this will lessen risk.

·2· Soaking the sliced fruit before cooking is not essential, but it is often a convenience as the fruit can be cut up and then left in water overnight. It also slightly reduces the initial cooking time.

·3· Be sure that the peel has boiled long enough for it to disintegrate when rubbed between finger and thumb before adding the sugar.

·4· In the second cooking (with the sugar) the larger the pan used, the less time the marmalade will take to reach setting point.

·5· It is always better to err on the side of undercooking. Overcooked marmalade is irretrievable, runny marmalade can be boiled up again.

·6· When the sugar has been added, let it dissolve completely before allowing the marmalade to boil. Once it has dissolved, boiling should be brisk but not furious.

·7· To test for setting drop a little of the juice on to a cold saucer and leave it to cool. If it jells and the surface wrinkles when pushed with the finger, the marmalade is ready to pot.

·8· Leave the preserve to cool for about 15 minutes and then stir well before putting into pots. This prevents the peel from rising to the top of the jars — which should be sterile and hot.

✳ *Lemon Marmalade* ✳

Wash 4 large lemons, cover them with cold water and leave them to simmer for about 2 hours. When they are soft enough to be penetrated by a matchstick, remove from the cooking liquor and slice the lemons into slivers.

Weigh the fruit, and for each pound of pulp add 1 pint of the cooking liquor and 2 lb preserving sugar. When the sugar has dissolved, boil briskly for about 20-25 minutes. Yield : about 3 lb.

✳ *Three-Fruit Marmalade* ✳

> 1 grapefruit
> 1 large sweet orange, 2 lemons
> 1½ pints water
> 2 lb preserving sugar

Squeeze the juice from the orange and lemons and cut the flesh from the grapefruit before slicing up the rinds of the fruit.

Tie all pith and pips into a piece of muslin and cut up the grapefruit flesh.

Cover the fruit with the water and simmer gently for about 2 hours. Remove the muslin bag, add sugar and boil briskly for 10-15 minutes after the sugar has dissolved. Yield : about 4½ lb.

* Grapefruit Marmalade *

> 3 grapefruits
> juice of 3 lemons
> 3½ pints water
> 3 lb preserving sugar

Cut the flesh from the grapefruit, and pare off any excess pith (tying it with the pips in a muslin bag). Slice the peel, cut up the flesh and, if it is convenient, leave both to soak overnight in half the water.

Next day add the rest of the water and simmer (with the muslin bag tied to the handle of the pan) for 1½-2 hours. When the fruit is soft, add the lemon juice and preserving sugar. Boil for 25-30 minutes after the sugar has dissolved. Yield : about 5 lb.

* Lime Marmalade *

A marvellously astringent breakfast preserve

Cover washed limes (it is not worth making with less than 10 or 12) with cold water, bring to the boil and simmer for 1½-2 hours until soft. Remove pips, slice the fruit thinly and weigh. Allow 2 lb sugar and 1 pint of the water in which they were boiled for each 1 lb of fruit. Boil fruit, sugar and water for 30 minutes till setting point is reached.

* Tangerine Marmalade *

> 4 lb tangerines
> 3 lemons
> 1½ pints water
> 4 lb sugar

Cut the fruit in half, remove all pips and slice finely. Cover with water and simmer till tender. Add the sugar. When it is all dissolved, boil briskly, about 20 minutes, till setting point is reached.

PRESERVES TO ACCOMPANY MEAT

Of all the preserves in the store cupboard none are more useful or more rewarding to make than a few off-beat accompaniments for cold meats, poultry and pies.

Certainly rowanberries, quinces and crab-apples, fruits of late sunshine and early frosts are treasure worth bribing the family to hunt for. And even city families are unlikely to have to hunt too hard. The suburbs of London are almost as well endowed with mountain ash as the Lakeland fells and the Scottish moors, and I have found crab-apples on Hampstead Heath, mulberries in Clapham and quinces in St. John's Wood.

∗ *Rowanberry Jelly* ∗

A delicious jelly can be made either with rowanberries on their own or combined in equal quantities with crab-apples.

Wash the berries and cut up the apples if used, cover them with water and simmer till soft. Strain through a jelly bag and measure the resulting juice. For each pint add 1 lb sugar and the juice of 1 small lemon. Bring slowly to the boil and then cook fairly briskly until setting point is reached — about 15-20 minutes. Pour into warm pots, but do not cover for at least 24 hours.

If crab-apples are used, omit the lemon juice.

∗ *Quince (or Pear) and Lemon Chutney* ∗

3 lb quinces or rather under-ripe pears
1 large lemon
1 onion, 2 cloves garlic
½ lb sultanas, ¼ lb stem ginger
1 pint white wine vinegar, 1 lb loaf sugar
1 tsp each allspice berries, coriander and salt
a few cloves, peppercorns, and blades of mace
a stick of cinnamon and a hunk of root ginger

Wrap the spices in a muslin bag, peel and core quinces or pears, chop lemon, removing the seeds. Put all the ingredients into a preserving pan. Bring slowly to boil and simmer until the fruit is soft, the volume reduced by more than half and the consistency rich and thick. This will take about 30 minutes for pears and 45 minutes for quinces.

* Blackberry Relish *

3 lb blackberries
1 lb Bramley cooking **apples**
¾ lb onions
1 pint cider vinegar
1 lb white sugar
½ oz ground gnger
1 tsp each, cinnamon, mace, allspice
6 cloves
6 peppercorns
2 tblsp gros sel or sea salt, or 1 tblsp ordinary salt

Peel and chop apples and onions and cook them with the black-berries, vinegar and spices for about 45 minutes. Rub through a nylon sieve, add the sugar.

When it dissolves, let the mixture boil quite briskly until it thickens, 10-20 minutes according to the size of the pan.

Pot immediately and cover with greaseproof paper and tie down with a piece of cotton dipped in melted paraffin wax, to prevent evaporation.

All spiced fruits are better for being left to mature a little if possible.

* Spiced Cranberries *

1 lb cranberries
2 gills cider vinegar
8 oz sugar
¼ oz each root ginger, stick cinnamon, allspice berries
4 cloves

Tie the spices together in a small piece of muslin. Put the cran-berries in a pan with the spice bag and vinegar, and simmer until the fruit is soft and the skins start to pop.

Add the sugar and boil gently for 20-30 minutes, by which time the mixture should be the consistency of strawberry jam. Remove the spice bag before turning the preserve into small warmed jars. Let it cool before sealing with paper and waxed cotton.

* Spiced Peaches *

1 lb peaches
½ lb granulated sugar
¾ gill cider or white wine vinegar
1 tsp each ground cinnamon, allspice, coriander
½ tsp ground cloves
¼ of a grated nutmeg

Mix the sugar and spices thoroughly before removing the skins of the peaches by plunging them into boiling water.   Cut them in half, cover first with sugar and then the vinegar.   Raise the heat very slowly until all the sugar is dissolved and boiling point reached.   Then simmer for 2-3 minutes, according to the size of the fruit, before transferring it (using a draining spoon) to wide-mouthed Kilner jars.   Continue boiling the vinegary syrup until it starts to thicken (5-10 minutes), then pour it over the peaches.   Seal the jars immediately.

Spiced peaches are superb with ham, gammon, pork, cold duck or pheasant.

* Sweet and Sour Cherries *

2 lb morello cherries
1½ pints white wine vinegar
¾ lb sugar
12 cloves

Leave ½-inch stalk on cherries.   Put them unstoned into wide-necked bottling jars.

Boil sugar, vinegar and cloves together for 10 minutes.   Pour over the cherries and seal.   Do not eat for at least a month.   Good with the dark meats — mutton, hare, venison — also pork.

* Apricot and Orange Relish *

There is no need to go hunting for the fruit for this relish.   Also it can be made all the year round because it actually tastes better made with dried apricots rather than fresh ones.

1 lb large dried apricots
½ pint white wine vinegar

1 lb Demerara sugar
1 orange
2 each chillies, cloves of garlic, shallots
½ oz preserved ginger
2 tsp gros sel or sea salt, or 1 tsp ordinary salt
½ dozen almonds

Cover the apricots with water and leave them to soak overnight. Next day, simmer the fruit until soft in the same water. Meanwhile, cut the chillies (all seeds removed), ginger, garlic and shallots into tiny slivers and put them all into a pan with the vinegar, salt, sugar and grated rind and juice of the orange.

As soon as the vinegar starts to boil, add the cooked apricots — cut in quarters. Let the mixture simmer for 20-30 minutes. Unless a really heavy pan is being used, the relish will need almost constant stirring. About 5 minutes before the cooking is finished, throw in the almonds, blanched and sliced.

* Green Tomato and Orange Relish *

If, by the autumn, you still have unripe tomatoes on your plants, they are likely to stay that way. So there is no loss in picking and pickling them.

This preserve, however, is a gentle alternative to pickling and it is an ideal partner for cold ham, pork or poultry.

1 lb of greeny-yellow tomatoes
¾ lb white sugar
2 oz crystallised ginger
2 oranges
2 lemons
allspice, cinnamon, coriander and ground juniper
3 tblsp white wine vinegar

Cut tomatoes into four, cover with sugar and vinegar and leave overnight. Next day, add all other ingredients (only the juice and grated skin of citrus fruits) and simmer till the tomatoes are translucent and soft. Spoon them into jars and then bubble the syrup until it shows signs of thickening. Wait till it cools before topping up the jars.

··· * ···

The important thing to remember about fruit in spirit is that tipsy fruits will taste as good as but no better than the spirit in which they are preserved.

* Peaches in Spirit *

It is conventional to assume that this spirit must be brandy, but recently I have found it more satisfactory to use vodka as the colouring in many brandies tends to stain the delicate complexion of peaches.

Start by making a light syrup with equal quantities of sugar and water (for 12 peaches allow about 1½ cups of each) brought to boiling point and kept there for 10 minutes.

Choose peaches the size of a de-fluffed tennis ball (4 should tip the scales just over the pound). Dip them (1 at a time) into boiling water for 10 seconds — the skin should then rub off readily. Make a circular incision round the peach, running the knife down the dimpled divide, and ease it in half with your fingers and remove the stone.

Place the halves in wide-necked jars. About 12 halves fit into a 2 lb Kilner jar, which is ideal for the purpose. Do not overfill the jars.

Pour the syrup over the fruit and leave it for 12 hours. Pour between ½ and ¾ of it off (keep this in a screw top jar until the next time you make a fruit salad) and top up the jar with vodka. Store for at least 2 months in a cool dark place. The same method can be used for preserving slightly under-ripe greengages or cherries, omitting the skinning process. Instead, prick the fruit 3 or 4 times with a bodkin.

* Brandied Prunes *

Carlsbad plums have vanished completely from the Christmas market and the dwindling supply of Elvas plums seems to get snapped up by those conscientious people who start their Christmas shopping in September. The rest of us have to fall back on home-preserved plums — or rather, prunes. And what better preservative than brandy?

> **prunes (enough to ¾ fill a 1 lb jar)**
> ½ gill water
> ¾ gill brandy (Australian will do)
> **2 oz sugar**
> **4 slices zest of orange**

Make a light orange syrup by boiling sugar, water and 2 slices of the orange peel for 5 minutes. Pack the prunes into a screw top jar interspersed with the rest of the zest, pour over the strained syrup and cover the prunes with brandy or whisky if you like it better. Seal tightly and shake the jar gently so that the syrup and brandy blend.

The longer the prunes have to mature, the better they will be.

-14-
HOLIDAY COOKING

First, and almost the only rule about country cottage and camping cookery is that there should not be much of it; also that there should be almost nothing to wash up afterwards. As there is rarely more than one burner (and that temperamental) in most caravans and cottages, the ideal ratio is one meal, one pot.

Camping is the time when tinned, tubed and packaged foods really come into their own, when it pays to invest in a miniature English cheese such as a double Gloucester, and a whole Continental sausage, a salami, for instance.

There are some packaged foods which even in the open air may not appeal as a meal in themselves but which provide an encouraging groundwork on which a good dish can be improvised. For instance, to one of those dry packaged *paellas*, onions, green peppers, celery, tomato or even a tin of prawns can be added.

A jar of Cirio tomato sauce can be used as it stands to dress spaghetti, or it could be improved with chopped chicken, olives, cooked mushrooms, what you will. Or it could be added to a fry-up of onions and minced beef.

The recipes following are all of the easy going variety to which extras can be added or some of the ingredients omitted, according to the convenience of the occasion, and they can all be cooked in one pot. If you face a fortnight of this sort of cooking and unless you are familiar with your temporary kitchen, it is worth taking along a sharp knife, a pair of scissors and a large, heavy pot suitable for either oven or stove top; also a roll of foil.   If you have an oven, even if it's only a tin box resting on top of a primus, cuts of pork, lamb, veal and fish can be seasoned, greased and wrapped up in foil.

* Veal and Tomato Casserole *

For 6 people
>   3 lb chopped veal (neck, shoulder or knuckle)
>   2 large onions
>   1 large tin tomatoes
>   3 tblsp olive oil
>   1 lb small scrubbed potatoes
>   seasonings, paprika and herbs as available

Heat the oil, fry the onions until they are golden, then do the same for the veal.   Sprinkle with paprika and stir thoroughly before pouring in the tomatoes.   Season with salt, pepper and herbs. Cover and cook slowly for about 2 hours.   Halfway through the cooking drop in some tiny potatoes, or some large ones cut small.

For a second showing add 2 small sliced green peppers and 4 oz mushrooms.   Just before serving, stir in 1 gill of cream soured with a few drops of lemon juice.

* Pork and Bean Casserole *

For 4 people
>   2 lb divided spareribs or 4 large lean pork chops
>   2 16 oz tins of baked beans
>   4 tomatoes
>   pepper and herbs, fresh or dried, basil for choice

Paint the surface of the meat with oil and let it brown lightly in a heavy pan before smothering it with the baked beans.   Add the chopped-up tomatoes, herbs and seasonings.   Cover tightly and leave to simmer slowly on top of the stove or in the oven for 30-45 minutes.   This can well be made in advance and reheated when it is needed.

* *Chicken with Corn off the Cob* *

For 3 people

> 1 packet frozen chicken drumsticks or 3 joints chicken or 1 small
>   jointed chicken
> 1 large tin of corn
> all or any of the following vegetables : mushrooms, spring
>   onions, peppers, courgettes, tomatoes
> 2 oz butter
> salt and pepper

Melt the butter in heavy pan and *sauté* the joints of chicken until they start to take colour. Add the sliced mushrooms and courgettes and toss them in the butter for a few moments before adding the tin of sweet corn.

Cut the peppers into slivers and roughly chop the tomatoes before stirring them into the mixture in the pan. Season, cover the pan and leave it to simmer slowly until the chicken is tender — about 45 minutes.

* *A Gratin of Crab Meat* *

Adapted from Marian Tracy's Casserole Cookery, *a wide-awake American cookery book now published in Britain.*

For 2 people

> 1 small tin of crab meat (7 oz approx.)
> 1 small tin mushrooms (3 oz approx.)
> 1 dessertsp each flour and butter
> $\frac{1}{2}$ gill each thin cream and cider or white wine
> salt, pepper, a pinch of dried mustard and a sprinkling of paprika
> breadcrumbs or a few slices of left-over potatoes

Tip the crab meat into a small casserole, fork it over to separate the fibres and mix in the mushrooms. Melt the butter in a small pan, remove it from the heat and stir in the flour.

Return to the fire for a few moments, stirring all the time, before slowly blending in the cream. When it starts to simmer add the cider, or wine, and all the seasonings except the paprika. Cook for a few minutes more before stirring the sauce into the crab meat.

Sprinkle the surface with breadcrumbs, a little paprika and dot with butter. Bake for 15-20 minutes in a preheated oven at Regulo 4 or 355 deg. F.

❋ *Instant Tomato Sauce* ❋

1 oz butter
1 large onion and 1 clove garlic
1 16 oz tin tomatoes
1 tblsp tomato purée
salt and pepper, any herbs you have to hand, and 1 lump sugar

Melt butter, colour the sliced onion, add the crushed garlic, tomatoes, tomato *purée* and seasonings. Simmer slowly in an open pan until most of the liquid from the tomatoes has evaporated and the ingredients have blended into a mixture the consistency of home made jam.

Serve with *pasta*, potato *gnocchi*, hardboiled eggs, meat balls, sausages, hamburgers or toasted cheese.

❋ *Egg and Bacon Sauce for Pasta* ❋

3 oz butter
4 oz bacon or ham cut into slivers the size of a matchstick
3 eggs
½ tsp black pepper, freshly ground if possible

Fry the bacon or ham in the melted butter, sprinkle with pepper and pour in the beaten eggs. As soon as the egg starts to set, stir the mixture into a dish of steaming hot spaghetti. The heat from the *pasta* will complete the cooking of the egg which should be set but not solid.

❋ *Rabbit and Cider* ❋

Soak a skinned and jointed rabbit in cider. Dry, then brown in oil. Add onion, tomatoes, mushrooms, herbs, bacon, seasoning and the cider used for soaking. Cover the pan and cook very gently for 1-1½ hours.

This can be made one day and served the next.

❋ *Kolokythia Soufflé* ❋

For 4 people
This *soufflé* from *Greek Cooking* by Joyce Stubbs is very much a peasant recipe, demanding neither special dishes nor regulated oven,

which makes it particularly suitable for holiday kitchens. Serve with cold meat.

    2 lb courgettes
    3 eggs
    4 oz butter
    6 oz Parmesan
    salt, pepper, nutmeg

Slice courgettes and simmer in salted water until tender. Mash them in a colander and leave to drain.

Mix in butter, grated cheese and egg yolks. Season. Beat egg whites separately and fold into mixture. Pour into well buttered dish, sprinkle top with more cheese and brown in oven.

* Spanish Rice *

For 3-4 people

    12 oz rice
    6 oz butter
    4 onions and 2 cloves garlic
    ¼ pint each cider and water, or just water
    2 tins tomato juice
    salt and pepper, herbs
    handful of grated cheese

Using a large frying pan cook chopped onions and garlic slowly till clear in 4 oz of the butter. Add rice, turn it over in the butter and as soon as it is shiny add the cider. Stir and let it cook slowly until ½ the liquid has been absorbed. Add herbs and seasoning.

Gradually add first the tomato juice, then as much of the water as is necessary. Never add more liquid until that remaining in the pan has been mostly absorbed, and stir at frequent intervals.

After about 40 minutes the rice should be cooked but the grains still separate, and there should be no liquid left in the pan. Stir in remaining butter and grated cheese and serve.

This is a good meal in itself, but even better if accompanied by a few slices of salami or ham.

* *Camp Chicken* *

For 6-8 people

Take a large boiling fowl and stuff it with mushrooms seasoned, spread with butter and touched, if you like, with garlic. Brown the bird in butter, add sliced carrots and onions and any herbs you can find. Brown these slightly before covering the bird with boiling water. Season with salt, pepper and mace. Simmer slowly for 1½ hours, then throw away the soggy vegetables and replace them with a selection of the following : carrots, onions, baby turnips, tomatoes, peas, broad beans, sprue asparagus, mushrooms, green peppers. Simmer for another ½ hour. Serve with dumplings made with 3 oz suet, 6 oz flour, a pinch of horseradish, salt and pepper, and enough water to bind. Drop these into the chicken liquor 15 minutes before the bird is ready to serve.

··· * ···

However primitive the cooking facilities, for most children a camp meal without a pudding is like a prologue without a tale.

First, then, the simplest sweet in the world :

* *Cream Cheese and Jam* *

Serve a slightly sweetened cream cheese with one of the more luxurious jams — cherry, raspberry or strawberry jam, or bramble or guava jelly. Home made is best, of course.

For most families fruit flans are the pick of the pops, especially when blackberries and bilberries are there for the picking. Although frozen pastry is fine for impromptu holiday pies, it cannot be kept without a fridge — a luxury I have yet to find in a cottage or caravan. So my next recipe is —

* *Pastry that Doesn't Need Rolling* *

> 6 tblsp soft butter
> 1 cup plain flour
> 1 tblsp castor sugar
> 2 tblsp cold water

Rub all the dry ingredients together until they are blended to the consistency of breadcrumbs, and then bind the mixture with water.

If possible, leave the paste to rest for ½ an hour before moulding it into a flan tin with the tips of the fingers.

* *Bilberry Flan* *

> 1½ cups bilberries
> 6 tblsp sugar
> 1 dessertsp flour
> 1 baked apple (optional)

Line a flan tin with the pastry. Mix the flour into the sugar before shaking it over the fruit. Cover the base of the tart with the flesh of the baked apple and fill up the tart with the fruit. Bake in a pre-heated oven at Regulo 6 or 405 deg. F. for 25-30 minutes.

* *Camp Treacle Pudding* *

This, the first of 2 impromptu hot puddings for chilly August days, is adapted from Florence White's collection of *Good Things in England*. It is a winner with children in spite of its inelegant name:

> 4 thick slices of white bread
> 4 generous tblsp each butter, golden syrup and sugar
> 1 gill milk

Cut the crusts off the bread and let it soak in the milk. Using a shallow frying pan, gently heat the treacle, sugar and butter until it turns brown and starts to give off a pleasant caramel smell. Put in the soaked bread and leave it in the syrup for 2 minutes. Serve with a slice of lemon or a little thin cream.

* *Spontaneous Sponge Pudding* *

> 1 packet sponge pudding mix
> ¾ lb any summer fruit (plums, blackberries, damsons)
> 2-4 tblsp sugar

Lay the fruit in any ovenproof pie dish. Sprinkle with sugar before covering the fruit with the sponge mixture — made up according to the directions on the packet. Bake for 45 minutes in a moderate oven, Regular 5 or 380 deg. F.

* *Summer Pudding made with Blackberries* *

Line a basin with thin slices of white bread fitted snugly together like
a jigsaw puzzle. Fill the cavity with blackberries which have been
lightly cooked with sugar but no water. Close the bread "box" with
a few more slices of bread, making sure that there are no gaps through
which the juice could escape.

Cover with a saucer, which should fit easily into the bowl, and
weight it down with anything heavy that lies to hand. Serve the
pudding next day with plenty of cream.

* *Spicy Gingerbread* *

Picnic cakes must keep well. Both this and the honey cake that
follows are, in fact, all the better for being made a few days before
they are needed.

$\frac{1}{2}$ lb each white flour and wholemeal flour
$\frac{1}{2}$ lb each golden syrup and black treacle
4 oz brown sugar
$\frac{1}{2}$ lb butter
3 eggs
1 generous gill of milk
2 oz each sultanas, candied peel and crystallised ginger
2 tsp ground ginger
1 tsp each cinnamon and allspice
pinch of nutmeg and salt
1$\frac{1}{2}$ tsp bicarbonate of soda

Prepare a 7 inch by 11 inch by 2 inch deep cake tin by painting it
with melted butter, lining it with greaseproof paper and then painting
the paper with more melted butter.

Put the syrup, treacle, butter and sugar to melt slowly in a heavy
pan. Meanwhile, sieve the flour into a bowl with the salt, spices and
bicarbonate of soda. The bits that remain in the sieve should be
tipped in at the end and well stirred. Add the sultanas and the
chopped ginger and peel.

Pour the milk into the well beaten eggs. Make a hole in the
middle of the flour and tip into it the sweet ingredients in the saucepan and the egg mixture. Work these gradually into the flour,
stirring from the centre outwards, until all the liquid is absorbed.
Pour the mixture into the prepared tin and bake for 1$\frac{1}{2}$-2 hours in
an oven preheated to Regulo 2 or 310 deg. F. Leave the cake to
cool in the tin.

✷ *Honey Cake* ✷

> 2 gills sour cream
> 1 egg
> 4 oz Barbados sugar
> 9 oz wholemeal flour
> 3 tblsp honey
> 1 oz each candied peel and sultanas soaked in 3 tblsp brandy or
>   orange juice
> 1 oz chopped walnuts
> 1 tsp baking powder

Whip cream, egg and sugar together, fold in flour, warmed honey, fruit with brandy or orange juice, and nuts. Mix very thoroughly before adding the baking powder. Turn the mixture into 2 buttered sandwich tins and bake in a preheated oven for 45 minutes at Regulo 3 or 335 deg. F. If a deep cake or loaf tin is used, allow 1½ hours at Regulo 2 or 310 deg. F.

✷ *Picnic Meat Roll* ✷

This meat roll recipe was given to me by Mrs. Allison of Dorrie's Cookshop in Edinburgh. She cooks it in a special open-ended meat roll jar made by Govancraft (the 2 lb size). Not having one, I used two ordinary cans from which I had cut both ends.

The roll can be served either hot with a tomato sauce or cold. Cold, it is ideal for picnics, accompanied by garlic bread and tomatoes.

> 1 lb minced steak or veal
> ½ lb minced bacon
> 6 oz soft breadcrumbs
> salt, pepper, nutmeg, mace
> 2 eggs
> 2 or 3 tblsp of strong stock (or meat juice left over from the
>   Sunday joint)

Work the meat, bread and spices together until perfectly blended. Add beaten eggs and sufficient stock to moisten. Put into lightly greased jar or tins, cover each end with greaseproof paper and steam for 2 hours.

* *Bacon and Egg Pie* *

Perfect for picnics

Line a pie plate with short pastry, cover it with crisply grilled back or streaky bacon cut into squares. Break 4, 5 or 6 eggs on to the bacon without breaking the yolks, cover with sliced mushrooms tossed in butter and season with plenty of freshly ground pepper and just a little salt.

Cover with pastry, seal the edges tightly and finish off as for veal and ham pie (*see p. 149*). Bake in a fairly hot oven, Regulo 7 or 425 deg. F., for 25 minutes. This is good both hot and cold.

* *Forfar Bridies* *

These are pasties filled with fillet steak, ideal eaten cold for picnics provided you don't forget the tube of French mustard.

THE PASTE *for each man-sized pasty*
 ¼ lb plain flour
 ½ oz lard and 1½ oz butter with salt
 just enough water to make a dough

THE FILLING
 2-3 oz shredded fillet steak
 1 tblsp each chopped beef fat (or suet) and onion
 parsley, salt, pepper and dried mustard

Roll the pastry into a thin square, cover one third with the meat sprinkled with chopped fat and onion and seasonings. Fold over the pastry, seal edges tightly, make an escape hole for the steam, glaze with egg yolk and salt and bake for 10 minutes at Regulo 7 or 425 deg. F. and 30 minutes more at Regulo 4 or 355 deg. F.

* *French Bread Sandwiches* *

·**1**· Spread a split French loaf with a savoury butter made by mashing 2 crushed anchovy fillets and a little parsley, chives and spring onion tops into 2 oz butter. Then season a thin flattened minute steak for each person with black pepper, paint with olive oil and fry quickly each side so that the meat is still slightly rare. For a picnic the bread can be taken along already spread, and the steak prepared on the spot over a camp fire.

·**2**· Split a French loaf, scoop out some of the soft crumb and replace it with the following mixture : 4 tomatoes, 1 onion, 1 tblsp capers, 1 small green pepper (de-cored and de-seeded), 1 oz each black olives and sweet gherkins, all finely chopped, bound with a little olive oil, and seasoned with lemon juice, pepper and a little salt. Press the loaf tightly together and chill.

∗ *Young Fisherman's Soup* ∗

Any youngster who actually lands a fish, or bags a rabbit at harvest time takes it for granted that the fruits of his chase will be gratefully received and turned into something delectable for dinner.

The easiest way I know of satisfying both the honour and the appetite of young fishermen is an old recipe reproduced by Dorothy Hartley in *Food in England* and called simply *Young Fisherman's Soup*.

> **1 lb assorted coarse fish (even the small bony ones are fine)**
> **2 each tomatoes, carrots, onions**
> **a bunch of sweet herbs**
> **1 gill wine vinegar**
> **soy sauce, salt and pepper**

Chop fish and vegetables coarsely, add herbs, vinegar and cover with water; bring to boil and after a few moments of rapid bubbling, reduce heat and simmer 1 hour. Strain, season and serve.

Should there, in fact, be an eel or two lurking in the creel, the cook's task is more rewarding. The only snag is killing and skinning the creature. Eels are as slippery as a soapy baby and have a tenacity for life surpassed only by the rich and ancient relative of fiction.

One infallible if not very orthodox way of coping with the problem is to grab the eel in both hands (gloved) and fling it to the ground as if in a rage. Stunned, it can then be finished off painlessly with a skewer in the spine where it joins the head.

Remove the skin by making an incision round the collar and easing the skin off as if it were a stocking. This is easier if you hang the eel up by putting a string round its neck. Cut off its head and clean as usual, before adding to the soup.

–15–
CHRISTMAS

In England we tackle Christmas with more gastronomic gusto than any other country in Europe, with the possible exception of Germany, but of all grand occasion dinners, Christmas dinner is surely the simplest. The menu is fixed, the embellishments traditional and the basic routine — roasting a bird and steaming a pudding — familiar to the plainest of plain English cooks. Traditionally the ritual roast on 25th December was a goose. Turkey, top of the bill at most of today's Christmas feasts, is a relative newcomer to our tables.

* *Roast Goose* *

Geese are at their best for roasting from 8-12 months after hatching. Choose a bird with a pliable bill and flabby feet, both bright yellow, and legs still touched with down. As with all poultry, a rigid breast bone is a sign of age.

Geese are extravagant creatures for the housewife as nearly $\frac{2}{3}$ of the bulk is bone. It is necessary to allow at least $\frac{3}{4}$ lb of goose for each serving. A 10 lb bird, weighed after dressing, should provide

an ample Christmas dinner for a family of 6, with enough left over
for a cold meal on Boxing Day.

Fresh, clean-tasting ingredients make the best stuffing for a goose.
Either the apricot or the prune and apple stuffing (*p. 248*) are suitable.

Having stuffed the bird (remember to allow room for the stuffing
to expand), prick it all over with a large needle and rub with salt
and lay the bird on the grid shelf of the oven, preheated to Regulo 8
or 445 deg. F. Place a baking tin containing a segment of orange
peel on the shelf immediately below the bird to catch the fat. After
about 12 minutes, when the fat has started to run, lower the heat to
Regulo 6 or 400 deg. F., and 15 minutes later reduce it again to
Regulo 4 or 355 deg. F. Baste every 20 minutes and allow a cooking
time of 20 minutes per pound, dressed weight. Half an hour before
serving season with black pepper and a little more salt.

∗ *Roast Turkey* ∗

In these fat-fearing days, turkey is the most popular, as well as the
most economical Christmas dinner. From the taste and texture
point of view smallish turkeys, weighing about 10 lb, make a better
meal than the vast farmyard giants, and a hen bird (poult) is said to
have more delicate flesh than the cock. A family of 6 should be
able to dine sumptuously twice at least off a 10 lb bird and still have
plenty over for sandwiches (the nicest way of eating turkey in my
opinion).

When choosing a turkey look for well rounded drumsticks, plump
breast, white skin and the usual pliable breastbone, and, if possible
of course, avoid a bird that has been deep frozen. One way of
adding an air of Christmas luxury to a turkey is to buy a tiny tin of
truffles and insert splinters of truffle under the skin of the bird at
least 12 hours before you intend to cook it. The most important
thing about turkey stuffings is that they should be moist and contain
enough fat material to counteract the natural dryness of turkey flesh.
Any of the stuffings in the section that follows are suitable for turkey.

Alternatively, a turkey can simply be stuffed with $\frac{1}{2}$-$\frac{3}{4}$ lb of fat
bacon rashers, a few sprigs of thyme or rosemary and a little butter.
Smear more butter on the breast, sprinkle lightly with black pepper
and cover with rashers of fat bacon. Envelop the bird completely
in a double layer of foil, folding it closely so that the fat and juices
cannot escape, but not too tightly. Put the bird in the oven preheated to Regulo 8 or 445 deg. F. and after 15 minutes turn the oven
down to Regulo 5 or 380 deg. F., and allow 25 minutes for each
pound of turkey as weighed when prepared for the oven, but before

stuffing. Twenty-five minutes before the cooking is complete fold back the foil covering to allow the bird to brown.

Serve with spiced cranberries (*see p. 227*).

* Walnut Stuffing *

½ lb each veal and lean pork
liver and heart of the turkey
¼ lb chopped walnuts
2 tblsp cognac
herbs and seasonings

Mince all the meats together and pound until well blended. Add roughly chopped nuts, brandy, parsley, thyme and seasoning.

* Mushroom Stuffing *

liver of the bird plus 3 or 4 chicken livers
¾ lb mushrooms, sliced and sautéed in butter
handful of fresh breadcrumbs
2 rashers of bacon
3 or 4 shallots and a clove of garlic
salt, pepper and parsley
2 eggs

Chop bacon and liver finely. Mix with mushrooms, hashed shallots, creamed garlic, parsley, seasoning and breadcrumbs. Bind with 2 beaten eggs.

* Chestnut and Sausage Stuffing *

1 lb chestnuts (with both skins removed, *see p. 165*)
1¾ lb sausage meat
whisky and seasoning

Cook the chestnuts in stock made from the giblets of the bird until they are soft, about 30-45 minutes. Fold them into the sausage meat. Add as much whisky as you feel inclined to afford (but not more than 4 or 5 tblsp) and a seasoning of crushed juniper berries, black pepper and salt.

* *An Easy Chestnut Stuffing* *

> 1 tin unsweetened chestnut purée
> 1 small tin pâté de foie
> 1 diced stick of celery
> 4 oz finely chopped bacon
> 6 oz mushrooms, sautéed in butter
> 1 chopped onion, also sautéed in butter
> the liver and heart of the bird
> parsley, salt and pepper
> a touch of garlic for those who like it

Work the *pâté* into the chestnut *purée* and stir in all the other ingredients.

* *Apricot Stuffing* *

> $\frac{3}{4}$ lb apricots
> 4 oz breadcrumbs
> 4 sticks celery
> 1 green pepper
> 1 apple
> 2 eggs
> 3 oz butter
> salt and pepper

Soak the apricots overnight, chop them up and combine them with the celery, pepper and apple, also finely chopped. Add breadcrumbs, beaten egg, melted butter and seasoning.

* *Prune and Apple Stuffing* *

> $2\frac{1}{2}$ oz boiled rice
> 4 sweet apples (peeled, cored and sliced)
> $1\frac{1}{2}$ dozen prunes (soaked, cooked, stoned and halved)
> the liver of the bird
> a stick of celery (chopped)
> the grated zest of $\frac{1}{2}$ a small lemon
> salt, pepper and ground mace
> parsley
> 1 egg
> 2 oz butter if it is going into a turkey

Combine all the ingredients, adding the beaten egg last of all.

Points about stuffing

·1· If possible, stuff the bird at least 12 hours before cooking to enable the flavours to permeate the flesh.

·2· Don't overstuff. The mixture will expand in the cooking. If you want to cut down on the cooking time of the turkey, stuff only the neck end, not the main carcase, as this slows down the cooking.

·3· If you haven't time to make any stuffing, fill the *turkey* with raw, chopped mushrooms marinaded in brandy or whisky; the *goose* with sweet apples, peeled, cored and quartered and similarly marinaded.

HAM ON THE BONE

Of all the gastronomic indulgences hallowed by the Christmas spirit, the one that gives me most pleasure is the almost Dickensian image of a whole dressed ham (or at least half of one) holding court on the sideboard. It has the added advantage of being the season's most sensible extravagance.

Pound for pound, ham is the least wasteful of the conventional Christmas meats (the ratio of flesh to bone is generous and every last scrap, even the fat and the bone, has a use), it is universally popular — unlike turkey, which some find insipid, and goose, which many find too fat — and it is something which, for a small premium, you can get the shop to cook for you.

There is also the pleasure, increasingly rare these days, of eating ham freshly carved from the bone.

Ham curing was a craft that was once as regional as cheese making. Some of the best of the local cures have survived, and top-class grocers such as Jacksons of Piccadilly, Paxton and Whitfield (93 Jermyn Street, S.W.1) and, of course, Harrods, take a pride in the variety of hams they display.

York, the most famous name in English ham, is dry salt cured. It has a pale rind and sweet, moist flesh that is rich without being strong. Its reputation rests not so much in having a monopoly of all the virtues one expects from a ham, as in the consistency with which it displays them.

Having a much wider market than some of the more *recherché* hams, it is slightly cheaper, though the branded Yorks (such as "Downes" and "Marsh and Baxter") are likely to be marginally dearer than the unbranded ones. But then you know you are getting what you paid for.

Bradenhams, the coal-black hams with the claret-coloured flesh, are unmistakable. This is the connoisseur's ham. The cure is sweet and mild, the process long enough to allow for months of maturing, and the flavour rich and sophisticated. Definitely a ham for adults and quite the most expensive, it would grace a Christmas buffet with all the magnificence of an Ethiopian Emperor.

Suffolk hams are marginally cheaper than *Bradenhams* and less dramatic in appearance, but they have a subtle flavour, sweetly strong and slightly aged, that I find every bit as appealing.

Worcester hams are synonymous these days with those cured by the Epicure Ham Company. Their No. 1 has a strong, full flavour reminiscent of a *Bradenham*; the No. 2 has the milder taste of a *York*. Neither quite reaches the standard of these aristocrats — but then neither do they fetch their prices.

Devon hams are the mildest and smallest of all, and, while not quite the cheapest per pound, mean the least outlay for the small family.

Irish hams are the cheapest, but buying them is a bit of a lucky dip. They can be excellent, but all too often the flavour is disappointing — either too fierce or too flat.

At time of writing the price of a ham varies from about 6s. a pound for those from Ireland to 12s. for *Bradenhams*. You can buy half a ham, or even a third if your grocer is accommodating, but, however much you buy, ask him to test it before you take it away.

This is done by running a knife into the flesh along the bone. If the ham is in perfect condition the knife will emerge smelling sweet and free of grease. Harrods do this as a matter of course.

Like many other grocers, they will also cook hams for customers provided they are given reasonable notice. They charge about 10s. for boiling and dressing with breadcrumbs and about £1 for sugar baking.

Most hams come with their cooking instructions attached — if so, follow these precisely, because there are subtle ways of emphasising a character of the cures. For instance, a little black treacle should be added to the water when boiling a *Bradenham* ham.

The general principles, however, are the same for all hams.

· 1 · Scrub under running water if the skin is rusty.

· 2 · Soak the ham in cold water overnight — rather longer if it is exceptionally dry looking.

· 3 · Wash it again and then put it into a pan deep enough for it to be completely covered with liquid.

· 4 · This may be water or a combination of water and cider. If you are cooking a smaller corner of ham or gammon which will be eaten up in 2 or 3 days you can add casserole vegetables such as carrots, onion, celery, etc., to the cooking water. But a ham you hope to keep for 2 or 3 weeks should not be cooked with vegetables.

• **5** • Bring slowly to boiling point but never let the water do more than simmer. If it boils, the flesh will harden. Skim as necessary.

• **6** • Allow 25 minutes' cooking time to the pound.

• **7** • To test whether the meat is cooked, pierce with a skewer.

: **8** • As soon as it is cool enough to handle, rip off the rind, then return the ham to the cooking liquor where it should stay until it is quite cold.

• **9** • Drain, pat it dry and dredge with lightly browned breadcrumbs. If you are going to the expense and trouble of buying and cooking your own ham, it is worth treating it to home prepared breadcrumbs rather than the sand coloured crumbs from a packet. Serve with Cumberland sauce (*see p. 187*).

∗ *Ayrshire Roll* ∗

Ayrshire bacon, rindless, mild cured and unsmoked, is a sweet and delicate meat; rolled, boiled, then baked and basted, it makes a magnificent hot dish for dinner or tempting cold cuts for lunch or supper parties.

Tie a 2 lb joint securely with string and leave it to soak for about 3 hours. Wrap it in a piece of butter muslin, cover with cold water and very slowly bring to simmering point (neither bacon nor ham should ever be allowed to boil). Allow half an hour's cooking time per pound, from the time the water starts to move.

As soon as the meat is cool enough to handle, remove the butter muslin and roll it in the following dressing :

> 3 tblsp fresh breadcrumbs, 1 tblsp **Demerara sugar**
> 1 tsp each dry mustard, mace and cinnamon
> 1 tblsp each sweet sherry or Madeira, and orange juice
> grated zest of an orange

Bake the roll in a preheated oven at Regulo 5 or 380 deg. F. for about 20 minutes, then spoon over it a mixture of melted butter (1 oz), brown sugar (2 tblsp) and cook for another 15 minutes.

If the meat is to be served cold — and I think it nicest that way — it deserves the compliment of something special to go with it, such as Cumberland sauce or spiced fruit (cherries, peaches, pears or crab apples), followed by a salad of onion rings, orange slices and watercress dressed with oil and a touch of lemon juice.

* *The Pudding* *

I have to admit I have never actually made a Christmas pudding, not even this one, though it has been the flaming glory of our Christmas lunch table for longer than I can remember, because this recipe, like the next one, is my mother's, and it is a family tradition which I have no wish to break that she should always make the pudding. I reciprocate by making the cake.

What distinguishes this particular pudding from all other puddings is not only the recipe, but the time allowed for maturing. Made in September for consumption this Christmas it will be a fine pudding, but made for the following Christmas it will be superb.

Of course making vintage puddings of this sort presupposes perfect storing conditions (cool and dry). Also the willingness to use more brandy than usual.

For most families several small puddings are more useful than a few large ones. Only schoolboys eat a lot of Christmas pudding at a sitting, so a small pudding always goes further than one would expect, and it's always worth putting up a couple of ½-pint sized puddings for the store cupboard.

> 1 lb fresh white breadcrumbs
> 4 oz flour
> 1 lb fresh chopped suet
> 2 oz freshly grated coconut
> 1½ lb stoned raisins
> 1 lb currants
> 1 lb sultanas
> ½ lb mixed candied peel (bought whole and chopped for choice)
> 3 oz almonds (blanched and chopped)
> grated zest of a lemon and an orange
> ½ lb Demerara sugar
> grated nutmeg
> 2 sherry glasses brandy (double this if the pudding is for next year)
> 1 pint barley wine
> 6 eggs
> 2 tblsp black treacle
> 1 grated carrot

Mix all the dry ingredients together, stir in the beaten eggs 1 at a time, and treacle. Add the barley wine and finally the brandy. Put the mixture into well greased basins in which a round of buttered greaseproof paper has been laid on the base.

Fill each basin about ¾ full, and press the mixture well down to prevent the formation of air bubbles. Cover with 2 or 3 more rounds of greaseproof paper and tie it down with a scalded square of linen or unbleached calico.

Steam for between 6 and 10 hours according to the size of the basin. Allow the puddings to cool, and the cloths to dry before storing them. On Christmas day they will need only 2-4 hours more steaming.

If you plan to keep the puddings until next year, replace the wet cloths immediately with dry ones.

Inspect the puddings once or twice during the year. If there is any sign of mould, remove the greaseproof paper, scrape away the mouldy pudding, put on new rounds of paper (soaked in cognac) and a new sterilised pudding cloth.

✳ *The Cake* ✳

If a cake is to live, it has to be a rich one, and rich cakes, like rich puddings, need time to mature, though not quite as long. A month in an airtight tin is adequate.

½ lb each butter and castor sugar
½ lb currants and sultanas
½ lb chopped mixed peel
2 oz each angelica and cherries (halved and floured)
½ lb self raising flour
½ tblsp mixed spice
grated lemon peel
a few drops of glycerine
4 eggs
almond essence

Line an 8-inch cake tin with a double layer of greaseproof paper and tie a piece of brown paper round the outside.

Cream the butter and sugar thoroughly before adding the well beaten eggs by degrees. If the mixture shows any sign of curdling, add just a little of the flour. Sift the spice and flour together and add them lightly to the mixture. Stir in all other ingredients.

Turn the mixture into the prepared tin and bake in a preheated oven for 4-5 hours, the first 2 hours at Regulo 2 or 310 deg. F., the rest of the time at Regulo 1 or 290 deg. F. Leave the cake in the tin for 8 hours after removing from oven.

Before putting it away in the tin to mature, puncture the cake with a knitting needle or skewer and drop in brandy with an eye dropper, using a sherry glass of brandy in all. Then wrap in foil.

* *Almond Icing* *

Enough for an 8- or 9-inch cake
>   1 lb ground almonds
>   ½ lb icing sugar
>   ½ lb castor sugar
>   juice of ½ lemon
>   2 eggs
>   2 tblsp brandy
>   a few drops of almond essence
>   2 tsp rose water

Sieve the sugars together, and work them into the ground almonds. Add beaten eggs, the liquids and flavourings. Knead till smooth.

* *Royal Icing* *

>   2 egg whites
>   1 lb icing sugar
>   1 tsp lemon juice

Agitate the egg whites with a fork, but do not beat them to a snow. Sift icing sugar twice, and then beat it, a little at a time, into the egg whites. When it is all absorbed add lemon juice and continue beating until the icing stands up in peaks. This will take longer than you expect.

If the icing is not to be used at once, cover with a damp cloth. The addition of ½ tsp of glycerine will make it easier to work.

Putting on the almond paste : Do this 48 hours before you apply the icing.

Level the top of the cake, dust away the crumbs and paint the cake all over with melted apricot jam. Sprinkle your pastry board with castor sugar and roll out the paste into a large circle about 12 inches in diameter. Place the cake upside down in the centre of this, and with your fingers and a jam jar press the paste evenly round the sides of the cake. Trim, and turn the cake the right way up again.

Icing the cake : Place it on an inverted dinner plate, and stand a palette knife in a jug of very hot water. Spoon a good dollop of icing on to the centre of the cake and, working from the middle out-wards, spread quickly over the top surface and just over the edge, roughing it with the edge of the knife. Beat the icing again for a moment before spreading it round the side of the cake with the reheated palette knife.

* The Mince Meat *

This is a Mrs. Beeton original. It appeared in the first edition of her book of household management, published in 1861. But for some inexplicable reason it was omitted by twentieth century editors. The recipe is reprinted here in the form in which it originally appeared :

Ingredients : 3 large lemons, 3 large apples, 1 lb of stoned raisins, 1 lb currants, 1 lb of suet, 2 lb moist sugar, 1 oz of sliced candied citron, 1 oz of sliced candied orange peel, and the same quantity of lemon peel, 1 teacupful of brandy, 2 tblsp of orange marmalade.

Mode : Grate the rinds of the lemons; squeeze out the juice, strain it, and boil the remainder of the lemons until tender enough to pulp or chop finely. Then add to this pulp the apples, which should be baked, and their skins and cores removed; put in the remaining ingredients one by one, and, as they are added, mix every-thing very thoroughly together. Put the mincemeat into a stone jar with a closely fitting lid, and in a fortnight it will be ready for use.

I find one large mince pie infinitely preferable to several small ones because, with the large pie, there is so much more mince than pastry, and who has room for pastry during the season of good will and self indulgence?

• WARMING MID-WINTER DRINKS •

In mid-winter a glass of hot, mildly spiced red wine, or a steaming mug of Farmer's Bishop is guaranteed to thaw the chilliest person — or party.

But really hot it must be, so use glasses with stems or handles to hold them by, and strong enough not to crack.

The only equipment needed for making a mull is a large and

scrupulously clean pan made of enamel, stainless steel or porcelain, and a ladle for dispensing.

Use a cheap wine (I generally use Algerian or Moroccan), spices in their whole form — rather than ground — and Seville oranges, if you have the choice, which is unlikely.

* *Mulled Wine* *

> 1 bottle cheap red wine
> 1 scant gill brandy (Australian, or French "pure grape")
> 1 gill water
> 10-12 lumps sugar
> 2 oranges
> 4 allspice berries, 4 cloves, 1 small stick of cinnamon

Impregnate the sugar with zest of orange by rubbing the lumps vigorously over the outside of the uncut fruit. Put the orange coloured sugar in a pan with the water and spices, bring it to the boil and leave to simmer for 5 minutes.

Add the wine, brandy, the juice of 1 orange and the remaining orange cut into horizontal slices. Heat slowly until the mull is too hot for the finger to bear, but is still well below boiling point.

Boiling drives off the alcohol and robs the brew of its potency — a point worth remembering if you are preparing mulled wine for a young people's party !

Another way of making the drink more innocuous (apart from the obvious one of weakening it with water) is to substitute Cherry Heering for all or part of the brandy, in which case a little less sugar should be used. A more spectacular trick is to burn off the alcohol just before serving.

Fireworks are also part of the fun of the next recipe :

* *Farmer's Bishop* *

> 1 gill water
> 1 orange
> 4 lumps sugar
> cinnamon, ginger, mace, allspice, cloves
> ½ gill whisky
> 3 gills cider (Whiteway's still vintage cider is excellent for this)

Cut the orange in half, stick it with 8 or 9 cloves and set it in a moderate oven to roast until the juice starts to run. Meanwhile simmer a small stick of cinnamon, a bit of root ginger, a blade of mace and ½ dozen allspice berries in the gill of water until it is reduced by three quarters of its original volume.

Put the roasted orange into a pan and add 2 tsp of the spicy water (put the rest into a tightly corked bottle, it will keep indefinitely and can be used for future mulls and punches).

Pour over the whisky, put a low heat under the pan and, as soon as the spirit is warm enough, set it alight. After 3 or 4 seconds, douse the flames with cider. Add the sugar and bring the Bishop almost to boiling point.

Oxford Bishop, a less rural drink, is made on the same principle. Instead of whisky and cider, use a medium sweet port and double the quantity of water. Flaming off the alcohol in both these Bishops is a matter of personal choice.

* Vin Chaud à l'Orange *

Cover 4 oz sugar and the thinly pared zest of 2 oranges with ¼ pint boiling water. Simmer gently for 5-10 minutes and then strain. Pour a bottle of cheap claret or Algerian wine into a stainless steel or porcelain pan, add the juice of the oranges and the syrup and heat almost to boiling point.

* Rum Punch *

⅔ pint tea (freshly made and stood for only 3 minutes)
½ pint orange juice
6 tblsp lemon juice
8 tblsp white sugar
½ pint rum
2 tblsp orange-based liqueur such as Cointreau

Pour the hot tea over the sugar, stir till it dissolves, then add juices and bring almost to boiling point before adding rum and Cointreau. Raise the heat once again, but do not boil.

* Mulled Ale *

Hot, mildly alcoholic and cheap

Bring 1 pint of mild draught beer nearly to boiling point, season it with grated nutmeg, a pinch each of powdered allspice, ginger,

S.T.C.B.—R

cinnamon and, if you have it, a touch of coriander. Add a strip of lemon zest (twisted to release the oil) and 1 tblsp of Demerara sugar. Just before serving add a port glass of brandy.

* Black Velvet *

A deceptively smooth mixture for igniting midmorning parties in the snowy season

Black velvet (or Bismarck) is simply a 50-50 blend of champagne and Guinness. Purists say that it is a ruination of good champagne and an insult to a worthy stout. In fact, it is an inspired partnership, as the champagne counteracts the slightly cloying quality of the stout, while the stout neutralises the acidity that hangs around the edge of all but the best champagnes (and no one would use vintage bubbly for black velvet).

What you need is something extra dry and costing not more than 25s. a bottle. Chill the bottles, but do not ice them, before pouring champagne into the stout. Have everything ready, but do not actually start the mixing until the first guest rings the door bell.

* Atholl Brose *

A traditional Hogmanay tipple in Scotland is Atholl brose (alias whisky broth). It is a more sustaining midnight drink than champagne and certainly a less predictable toast. It is also much cheaper.

Cover 2 cups of oatmeal with 4 cups of water. Leave it to stand overnight, then strain it through a nylon or stainless steel sieve, pressing the oats firmly to extract all the liquid.

Measure the liquor and for each cup stir in 2 tblsp each of warmed clear heather honey and double cream, before adding ¾ cup Scotch whisky. Atholl brose can be made some time before the party, but must be well stirred before it is served.

• TWELFTH NIGHT •

Twelfth Night, or the Day of the Kings, is the almost forgotten feast celebrating the arrival of the Magi in Bethlehem. Still observed by the Church as the religious festival of Epiphany, it is no longer the occasion for wassailing, bonfires and all the blatantly pagan customs that enlivened Twelfth Night up to the middle of the last century.

These days, January 6th is little more than the deadline for ripping down the dessicated greenery and throwing out the Christmas tree, but two Twelfth Night specialities are too good to be allowed to disappear into oblivion : Twelfth Night cake (what the French call *Gateau des Rois*) and *lamb's wool*. The spicy cake and steaming wassail marry wonderfully well and would make excellent fare for a late evening party.

✱ *Twelfth Night Cake* ✱

$\frac{1}{2}$ lb each butter, sugar and plain flour
5 eggs
1 lb currants
2 oz almonds, blanched and chopped
juice of 1 orange and the grated zest of 1 lemon and 1 orange
$\frac{1}{4}$ gill brandy
pinch each of allspice, cinnamon, mace, ginger and coriander (all
    ground)

Prepare a 7-inch cake tin by lining both bottom and sides with a double thickness of buttered greaseproof paper.

Soften the butter and beat in the sugar. Continue beating for several minutes, until the mixture is smooth and fluffy. Sift the flour and spices together, and then add a little of the flour to the butter mixture before adding the first egg. Stir thoroughly before adding a little more flour and another egg.

Continue in this way until all the eggs and flour have been absorbed. Fold in the prepared fruit and almonds, the grated zest of lemon and orange, and finally the brandy and orange juice.

Put half the cake mixture into the prepared tin, and lay on it, slightly off centre, a haricot bean wrapped in greaseproof paper. (Whoever gets the bean when the cake is cut is automatically elected King of the Bean — the legendary hero of Twelfth Night revels.)

Pour in the rest of the mixture and make a tiny dimple on the surface of the cake to encourage it to rise evenly.

Bake in the centre of a preheated oven set at Regulo 4 or 355 deg. F. for 3-3$\frac{1}{2}$ hours. After the first hour turn the oven down to Regulo 2 or 310 deg. F., but resist the temptation to open the oven to see how the cake is getting on.

Make sure that the cake is properly cooked by testing it with a clean skewer before removing it from the heat. Let the cake cool in its tin.

* *Orange Water Icing* *

If it is the intention to wash down this cake with *lamb's wool* or any other wassail, it should not be iced. But for any other occasion a light water icing would give it a more festive air.

>¾ lb icing sugar
>3 tblsp of orange juice

Add the juice, a teaspoonful at a time, to the sifted icing sugar, beating vigorously all the while. Continue beating for 5 minutes after the juice has been absorbed.

When it is smooth and malleable, pour the icing on to the centre of the cake and assist it to run evenly over the whole surface of the cake with a palette knife dipped in boiling water.

* *Lamb's Wool* *

The curious name *lamb's wool* is said to be a corruption of La Masubal, the patron saint of fruits and seeds; but I am more inclined to believe that it owes its name to the woolly apple pulp that is one of its most essential ingredients.

Whatever the derivation, *lamb's wool* is a warming and delicious brew, easily made and not expensive.

Serve it in small tumblers or sturdy wineglasses, unless you happen to own a set of punch glasses.

Enough for 12 glasses
>1 pint ale
>¾ pint white wine (ideally a Graves, though I suspect that originally lamb's wool was made with fermented apple juice)
>6 tblsp brown sugar
>seasoning of cinnamon, ginger and nutmeg
>4 small apples

Bake the apples until the flesh fluffs up. Heat the ale and wine together and stir in the sugar and spices. Put the cooked apples through a sieve and dilute the pulp with a little of the hot ale before stirring it into the main body of the liquid.

Bring it all back nearly to boiling point and serve immediately. To make the whole thing look rather splendid, serve it from a punch bowl, with roasted apples floating on the surface.

WINE

-16-
WINE

"A meal without wine is like a day without sunshine" says the slogan
across the foot of the menu in one of London's smaller French
restaurants. The British, like most North Europeans, are as used to
meals without wine, as they are to days without sunshine. But since
the mid-1950s many more have been travelling far and wide in search
of the sun, and finding and getting to know the wine as well, for
naturally it is in the sunny countries that the grape vine flourishes
and is at its most prolific.

Abroad, the obvious wine to drink with one's meal is the local
wine, because it is cheaper for one thing, but also because over the
years the people of the district have taken care in many cases to
match their wine to their food. It has been said that "Nature
produces the grape; man makes the wine". In other words, there
are many ways of modifying the fermented juice of the grape, even
though that very juice may have an innate chemistry derived from
its variety, the stock on which it grew, the soil and climate and the
treatment of both plant and fruit.

This means that to acquire any deep knowledge of wine and
wine-making calls for long and even profound study to which this
writer makes no claim. Our concern in this short chapter is to

dispense a little modest advice aimed at simplifying the use of wine at the table, and incidentally dispelling a number of minor misconceptions.

To quote André Simon, "Simple wines, young and inexpensive, should be our daily wines : they will make our meals so much more interesting and enjoyable, more beneficial also, since it is the food that we enjoy most that is sure to do us the most good". It's a nice afterthought and, if you like, a good excuse, but the quote was lifted mainly for that polite word "inexpensive".

The truth of the matter is that enjoyable wine need not cost crippling sums, but it takes a little bit of thought to hit on precisely satisfactory kinds of wine to serve to family and friends because people have such varied tastes. Some never touch white wine, others drink nothing else. But according to a recent survey more men in Britain prefer red wine, compared with women who lean towards more individual tastes, though they hate to be suspected of having a preference for softer, sweeter wines (even when they do).

This red-white division is sometimes bridged with pink wine, usually known by its French title *rosé*, but *la vie* entirely *en rose* could be monotonous, especially when I tell you that some Spanish *rosados* are merely a mixture of cheap reds and whites. The real *rosé* is made by gently pressing the white juice out of red grapes, and leaving the unfermented juice briefly in contact with the skins so that it acquires a pink tinge and a slightly fuller flavour. Famous French pinks are Tavel from Provence, Anjou from the Loire (generally sweetish) and *pelure d'oignon* or "onion skin", presumably because it has a russet tinge. Semi-sparkling or *pétillant* pink wines from Portugal are popular now in Britain, especially Mateus. And from Italy there is another, almost red, Chiaretto del Garda.

These are not what I would call serious wines, though they are pleasant and enjoyable, and they do nothing to resolve on a sophisticated level that old controversy about what to drink with what. André Simon, in his essay *Partners*, parenthetically excluding Champagne from his list of what goes best with what because it goes with everything, says no book can tell us, so much depends upon the mood of the moment, the company, the weather.

Taste counts, of course, but chemistry plays a small part, for the tannin in red wines has a tendency to set up in the mouth a not too palatable liaison with the phosphorus in some strongly flavoured fish. So, on the whole, a white wine devoid of tannin and also of sugar ("dry" as the jargon has it) is generally preferred with most fish. It has been said that lampreys are probably the only fish with which you should drink red wine. It has also been said, by Raymond

Postgate, that the sensible wine drinker drinks what he likes, though I am sure that he did not mean that we should ignore the preferences of our friends. Certainly experience will teach the discriminating wine drinker that there are right and not so right combinations of tastes. If the cook may be allowed to recommend a sauce for a certain meat or fish, let the wine lover recommend a wine to go with both, without being classified as a Wine Bore or a Wine Snob.

The first lesson is that with *hors d'oeuvre* that may be fishy, white wines free from sulphur and sugar are ideal. To quote prices would be folly, with surcharges and duties going up and down (but mostly up), so it would be wisest merely to mention that the driest whites are those of Burgundy and Chablis, Alsace, Moselle, Muscadet and Bordeaux. The last is tricky, for so many white Bordeaux are sweet and semi-sweet, especially the ubiquitous Entre-deux-mers. Be aware, too, that Chablis is seldom cheap when genuine and never genuine when it is a bargain. Petit Chablis, gathered from a wider area, is often much cheaper, but expect less of it in terms of fragrance and fullness of taste.

Sherry and Madeira can sometimes be used as table wines, but again experiment is perhaps the only way to be sure that the combination of flavours is satisfying. A really dry sherry such as a natural *fino* (pale and unsugared) or Manzanilla can be drunk with prawns. These two, and dry Madeira (*sercial*), also make a most pleasing accompaniment to soup of certain kinds, even fishy soups like *bisque* of lobster.

With savoury starters any humble red will do, provided that it is not so full-bodied and coarse that the uninitiated are repelled. A meal without fish can be happily accompanied by any ordinary red wine, and these days there are so many to choose from (apart from the classic areas of Bordeaux, Burgundy and the Côtes du Rhône) within quite reasonable price brackets that it should be possible to please most palates and pockets.

Before launching warily and precariously upon the dangerous project of advising other people about their wine, I should say that if you are fortunate enough to know an intelligent wine merchant, rely on his advice. He will have tasted, and probably drunk, more wine than you may ever be able to afford. Talk to him, listen to his answers, try his wares, tell him the truth (as you see it) about them. Take his advice about what to lay down and what to drink, and you will not need mine. This little essay is confined to hints for those who do not have a wise and kindly wineman.

First, cheap red wines are intended to be drunk young and fresh as you would drink them where they were made. Keeping them will do little or nothing for them by way of maturity; some may even deteriorate. Beaujolais, of which it is said more is drunk in Paris alone than was ever grown, is the leading instance. In the Beaujolais area it is even customary to drink the wine of the year as soon as it is ready. The keeping qualities of red wines depend upon their chemical make-up, so that the "harder" the wine, generally, the longer it will take to "come round", in the delightful language of the trade. This hardness comes from natural tannin, which is at its highest in wines made from grapes that have not been de-stalked. To hasten the maturity of red wines the stalks are removed before pressing. An excess of tannin may produce a hard wine (which then has when young the same sensation in the mouth as a strong cup of tea without milk) which takes thirty years to come round, or possibly never does.

This characteristic belongs almost exclusively to the clarets, the red wines of Bordeaux, where there are at least 50,000 growers, and several thousand distinguishable wines, including some of the world's most distinguished : the classified *château* products of the Médoc (the leading four, Lafite, Latour, Margaux and Mouton-Rothschild are beyond reach for most of us); the famous red Graves, Haut-Brion; the star turn of the Sauternes, Yquem. But there are thousands of less celebrated wines of reasonable quality, including the so-called *bourgeois*, artisan or peasant growths (a nice class distinction, the superior *crus* being implicitly noble) that fetch quite low prices, being unknown to the majority of wine drinkers, simply because they have not bothered to study "form", something they would certainly do with horses or football teams.

These minor clarets, together with the commune wines (blends from the larger subdivisions bearing names like St. Julien or Listrac), Médoc or simply Bordeaux *rouge* all partake of similarities of character, though the wider the area from which the blend comes the fainter the discernible qualities. At the extreme of cheapness and ordinariness come the *ordinaires*, the wines that the French drink every day, compounded of Heaven knows what proportions of poor quality Bordeaux, nondescript Midi wines and the fuller bodied North African varieties. Such wines are now sold in Britain by the large chains under Gallic brand names, and the only way to choose them is to try them, preferably with food.

Simple rules of thumb, arrived at over generations of wine drinking and not arising from selfconscious winemanship (a book as yet unwritten by Stephen Potter) may be sketched thus. On the whole serve white before red (if you have more than one wine that is) and dry before sweet. Champagne is once again the honoured exception that helps to prove the rule, for a sweet Champagne at the beginning of a meal could be The End.

The light wine precedes the heavy one, the young precedes the old, the humble before the well-bred. A French authority, though according to British oenophils the French are careless about their finest product, argues that two red wines of different vineyards should not be brought to the table during the same meal. Rather serve two from the same vineyard, younger before older, making an interesting and instructive contrast.

This same sensitive commentator protests at the habit of serving the pudding with dry Champagne. The sweet food shows up the acidity of the wine, the very acidity that gives the freshness to the taste when savoury flavours or none at all are present. Better to serve no wine at all at that stage of the meal. In any event, three wines are enough, especially when apéritifs or sherry have been taken beforehand. To quote Louis Jacquelin and René Poulain in their *Wines and Vineyards of France* :

"But certain menus demand four wines and this again is sensible and normal. Above this amount one is satiated. The sense of smell and of taste disappear. One is no longer drinking; one merely takes in alcohol without discernment or pleasure."

Balancing the tastes of food and wine may seem to earthier temperaments an unnecessary and extravagant adventure in gluttony. But if you have a palate and training it amuses you, just as training the eye or ear to take pleasure in pictures or passacaglias amuses others, then shrug off this guilty feeling, get hold of a wine list and plan a dinner party. But whatever the ascetic may say, he cannot dispute the fact that to drink good wine with discrimination and care is better than hitting the nerve centres with hard liquor. That amiable gourmet Cyril Ray believes that spirits before a meal with wine will numb the palate and rob the diner of some of the pleasure. Amen to that, but *chacun à son goût* and a "mouth" the morning after to the rest.

You will notice that I have focussed on French wines so far. That is because most of our better wines for the table (as opposed to the fortified wines) are French. A great deal comes from Spain, of course, but most of that is sherry. While France sends more than eight million gallons of wine a year, Italy sends only two and a half million. Of her wines Chianti is the most famous and the most popular in Britain, but names like Valpolicella and Bardolino, Soave and Barbera are becoming more widely known. It is pleasing to consider that Italian wines go best with Italian food (on the grounds that this is the way the makers drink them), and this is often so. A strange, sweetish sparkling wine from Emilia, Lambrusco, was once served to me with quails in a works canteen in Northern Italy. It seemed appropriate there and then.

Germany's wines are among the noblest made, grown as they are in some of Europe's most northerly vineyards. They are mostly white and come from the Rhine, the Moselle and from Franconia. Their immense variety and their complex pedigrees when they are great have had two results for the British hock fancier. First, only two names have made much impact, Schloss Johannisberger from the Rheingau, and Berncasteler Doktor from the Moselle. Secondly, like J.B. and Roscoe of Blue Nun fame, baffled by names like Niersteiner Rohr-Rehbach Riesling und Sylvaner Trockbeerenauslese, we ask for Liebfraumilch. Hard words have been said about this notorious name, just as they have about Beaujolais. It has a definition in German wine law which says that it may be applied to any wine of reasonable quality grown in the Rhineland. Lawyers have grown fat on laws like that. And the contents behind such a label remain an enigma, even to quite expert wine buyers. The name seems to have derived from the Church of Our Lady (Liebfrauenkirche) at Worms, and the Liebfrauenstifte vineyards there still produce a remarkable and beautiful wine.

But Liebfraumilch itself has become a convenient tag for blends whose character can be made consistent so that uninstructed hock drinkers can be sure of getting just that familiar bouquet and taste that they like, year after year, in spite of vagaries of suns and seasons. Beware of imitations would be an empty injunction; they are many. Know your merchant, know your shipper and try them out for yourself, is the best that one can do by way of advice. Liebfraumilch formulae are as secret as strongbox combinations.

The rising price of wine has attracted into Britain hordes of hitherto little known wines. Well established, and deservedly, are the Spanish and Portuguese reds, notably Rioja and Valdepeñas from the one, Dão and Vila Real from the other; the Rieslings (made from the same grape as most hocks) from Yugoslavia and Hungary; and Chilean Cabernet, made from the most used claret grape. There are others from North Africa, Greece, Russia and the Commonwealth countries, even from Turkey and Rumania. On the whole the cheaper reds are safer to experiment among than the cheaper whites, which tend to be acid and even sour.

About these there is little or no general advice to be offered, except "Try them". They will not cost much to experiment with and one or two may reveal themselves as acceptable and economical substitutes for the finer wines that cannot be afforded every day.

There are three other vital matters to be mentioned.

STORING

Laying down wine means tying up money (not all stored wine is a good investment) so, on the whole, this is best left to your merchant who can store your chosen bottles at proper temperatures and watch them for signs of maturity or deterioration. To have your own "cellar" under the stairs or in a corner is fun, and convenient too, but store only the most robust of reds (the kind that British Railways carry on their trains, for example) unless you have a coolish place where the temperature does not fluctuate too much. Storing white wines except for fairly immediate use is less wise. Their life is short and you could let them go "over the top", when they darken and lose their freshness. Port is a subject on its own. Amateurs should spend a little time on it before they spend any money.

SERVING

Serving wine is simpler than is generally supposed. Young reds and whites should simply be poured out of the bottle. A decanter is either pretentious or a pretence. True, a rich red Moroccan wine looks well through clear glass. The decanting basket or cradle is useless except as a means of transport for wines that have thrown a sediment while lying on the rack with their labels uppermost. Lay them gently in their cradle, as gently draw the cork (preferably with a double twist corkscrew) and pour the wine into a decanter, with a light behind the shoulder or the neck of the bottle, until you see the first threads of sediment stealing along. STOP, and leave the dregs for the cook — not to drink, but to add to the gravy. How do you tell which wines have thrown a sediment ? Experience and common sense. On the whole, it is the older, fuller bodied wines that deposit sediment. Occasionally one meets a Beaujolais or even a sherry that has done this. The sherry bottle being clear, there is no mystery. With dark green bottles, pour carefully.

Few wines improve if left standing about. They should lie down with the wine touching the porous cork. Once opened the majority of wines have a short useful life. Generally, the sweeter the wine, the longer it will keep, though this does not apply to port. Sweet

Tokay, for instance, Sauternes and Barsac, and the costly sweet hocks (made from those *trocken*, dried, *beeren*, grapes, *auslese*, specially chosen) will take a little oxygen for a few days or even weeks. But better not.

Further, to open a delicate, mature wine too long before a meal may "kill" it, that is disperse the attractive aroma and oxidise away some of the taste. Miss Iris Murdoch in *A Severed Head* gives her wine merchant hero the line :

"How many times must I tell you never to drink claret unless it has been open at least three hours?"

It does not do to be so dogmatic. Some wines may need three hours (but surely not if decanted); for others it would be a death sentence. For most, fifteen minutes to half an hour. Trial and error, mostly error, will show. Some old wines need decanting to rid them of that initial mustiness sometimes called "bottle stink".

TEMPERATURES

As to temperatures; cool, not ice cold, for whites of quality (freezing the cheaper ones helps them along); room temperature, unless you are an American, for reds, that is about 60-65 deg. F. (about 15-18 deg. C.). I was once served a steaming Beaujolais, because the label said "Serve at room temperature" and the room was about 80 deg. F. Incidentally, Beaujolais and a few other reds are often pleasanter cool, as is very dry sherry. It is all a matter of taste, but some tastes are more equal to the occasion than others.

A last word for the timid on the matter of what to buy and where. There are a number of wine societies whose tasting committees are well versed in seeking out and stocking up with good wines when the price is low. The most venerable is the International Exhibition Co-operative Wine Society (8-10 Bulstrode Street, London W.1), founded 1867, with a long and varied list. Entrance requires one sponsor, a member of course, and a £5 share that is returnable on request. Part of the International Distillers and Vintners group is the Directors Wine Club (37 New Bond Street, W.1, 21s. a year), with a short list chosen by, among others, Cyril Ray and David Peppercorn of Morgan Furze.

The Private Wine Buyers' Society (109/110 Jermyn Street, S.W.1, 21s. a year) goes in for buying by the cask and bottle it yourself, an exercise which demands a chapter to itself but is not going to get one. There are several others, but these three will take care of the vinous needs of most people. The purport of the message is, rely on others or educate yourself enough to be self reliant within limits.

The moral of the whole story is that wine drinking, like any other acquired taste, consumes time and money. Whether it is worth the effort and expenditure is a personal decision. But, if you decide in favour of taking an interest, you will be joining a fellowship that goes back before Homer and is as much of a bond between its members as politics or sport.

INDEX

INDEX

INDEX

275